THE DESULFURIZATION OF HEAVY OILS AND RESIDUA

CHEMICAL INDUSTRIES

A Series of Reference Books and Text Books

Consulting Editor

HEINZ HEINEMANN

Heinz Heinemann, Inc.,
Berkeley, California

Volume 1: Fluid Catalytic Cracking with Zeolite Catalysts, *Paul B. Venuto and E. Thomas Habib, Jr.*

Volume 2: Ethylene: Keystone to the Petrochemical Industry, *Ludwig Kniel, Olaf Winter, and Karl Stork*

Volume 3: The Chemistry and Technology of Petroleum, *James G. Speight*

Volume 4: The Desulfurization of Heavy Oils and Residua, *James G. Speight*

Additional Volumes in Preparation

THE DESULFURIZATION OF HEAVY OILS AND RESIDUA

James G. Speight
Corporate Research Laboratories
Exxon Research and Engineering Company
Linden, New Jersey

MARCEL DEKKER, INC. NEW YORK AND BASEL

Library of Congress Cataloging in Publication Data

Speight, J. G.
The desulfurization of heavy oils and residua.

(Chemical industries ; v. 4)
Includes bibliographical references and index.
1. Petroleum--Refining--Desulfurization. I. Title.
II. Series.
TP690.45.S628 665.5'34 81-7825
ISBN 0-8247-1506-3 AACR2

MARCEL DEKKER, INC.

270 Madison Avenue, New York, New York 10016

Current printing (last digit):
10 9 8 7 6 5 4 3 2 1

PRINTED IN THE UNITED STATES OF AMERICA

PREFACE

With the inception of hydrogenation as a process by which both coal and petroleum could be converted into lighter products, it was also recognized that hydrogenation would be effective for the simultaneous removal of nitrogen, oxygen, and sulfur compounds from the feedstock. However, with respect to the prevailing context of fuel industries, hydrogenation seemed to be not economical for application to petroleum fractions. At least two factors dampened interest, (1) the high cost of hydrogen and (2) the adequacy of current practices for meeting the demand for low-sulfur products by refining low-sulfur crude oils, or even by alternate desulfurization techniques.

Nevertheless, it became evident that reforming processes instituted in many refineries were providing substantial quantities of by-product hydrogen, enough to tip the economic balance in favor of hydrodesulfurization processes. In fact, the need for such commercial operations has become more acute because of a shift in supply trends that has increased the amount of high-sulfur crude oils employed as refinery feedstocks.

There are several valid reasons for removing sulfur from petroleum fractions; these include: (1) reduction, or elimination, of corrosion during refining, handling, or use of the various products; (2) production of products having an acceptable odor; (3) increasing the performance (and stability) of gasolines; (4) decreasing smoke formation in kerosenes; and (5) reduction of sulfur content in other fuel oils to a level that improves burning characteristics and is environmentally acceptable. In order to accomplish sulfur removal, use is still made of extraction and chemical treatment of various petroleum fractions as a means of removing certain sulfur types from petroleum products, but hydrodesulfurization is the only method generally applicable to the removal of all types of sulfur compounds.

Overall, we have seen emerge a growing dependence on high-sulfur heavier oils and residua as a result of continuing increases in the prices of the conventional crude oils coupled with the decreasing availability of these crude oils through the depletion of reserves in the various parts of the world. Furthermore, the ever growing tendency to convert as much as possible of lower grade feedstocks to liquid products is causing an increase in the total sulfur content in refined products. Refiners must, therefore, continue to remove substantial portions of sulfur from the lighter products, but residua and the heavier crudes pose a particularly difficult problem. Indeed, it is now clear that there are other problems involved in the processing of the heavier feedstocks and that these heavier feedstocks, which are gradually emerging as the liquid fuel supply of the future, need special attention.

The hydrodesulfurization of petroleum fractions has long been an integral part of refining operations, and in one form or another, hydrodesulfurization is practiced in every modern refinery. The process is accomplished by the catalytic reaction of hydrogen with the organic sulfur compounds in the feedstock to produce hydrogen sulfide, which can be separated readily from the liquid (or gaseous) hydrocarbon products. The technology of the hydrodesulfurization process is well established, and petroleum feedstocks of every conceivable molecular-weight range can be treated to remove sulfur. Thus, it is not surprising that an extensive knowledge of hydrodesulfurization has been acquired along with development of the process over the last few decades. However, most of the available information pertaining to the hydrodesulfurization process has been obtained with the lighter and more easily desulfurized petroleum fractions; but it is, to some degree, applicable to the hydrodesulfurization of the heavier feedstocks such as the heavy oils and residua. On the other hand, the processing of the heavy oils and residua present several problems that are not found with distillate processing and which require process modifications to meet the special requirements that are necessary for heavy feedstock desulfurization.

It is, therefore, the object of the present text to indicate how the heavier feedstocks may, in the light of current technology, be treated to remove sulfur and, at the same time, afford maximum yields of low-sulfur liquid products.

This text is designed for those scientists and engineers who wish to be introduced to desulfurization concepts and technology as well as those scientists and engineers who wish to make more detailed studies of how desulfurization might be accomplished.

Chapters relating to the composition and evaluation of heavy oils and residua are regarded as being necessary for a basic understanding of the types of feedstock that will necessarily need desulfurization treatment.

For those readers requiring an in-depth theoretical treatment, a discussion (Chapter 3) of the chemistry and physics of the desulfurization process has been included. Attention is also given to the concept of desulfurization during the more conventional refinery operations (Chapter 4).

However, no attempt is made, at any point in the present text, to enter upon discussions related to the removal of sulfur from organic materials by the various chemical methods, such as Raney nickel desulfurization. Adequate reviews of this particular subject are available, and the minimal relevance of this aspect of desulfurization to the current text did not appear to warrant the inclusion of such a section here. Nor, for that matter, is any reference made to the desulfurization of petroleum fractions by treating in a "hydrogen-donor" solvent—the process itself is well known (and mature) but the commercialization (with respect to desulfurization as part of a refinery operation) has, as yet, to be realized.

The effects of reactor type, process variables and feedstock type, catalysts, and feedstock composition (Chapters 5 and 6) on the desulfurization process provide a significant cluster of topics through which to convey the many complexities of the process. In the concluding chapters examples, and brief descriptions, of commercial processes are presented (Chapter 7) and, of necessity, some indications of methods of hydrogen production (Chapter 8) are also included.

Finally, the author wishes to note that this work was completed while he was an Associate Research Officer at the Alberta Research Council, Edmonton, Alberta, Canada. In this respect the author acknowledges the valuable assistance of Mrs. Tara Emter during the preparation of the manuscript.

James G. Speight

CONTENTS

THE DESULFURIZATION OF HEAVY OILS AND RESIDUA

1

THE COMPOSITION OF HEAVY OILS AND RESIDUA

1-1 THE PHYSICAL COMPOSITION OF HEAVY OILS AND RESIDUA

The definition of a heavy oil is quite arbitrary, but the term heavy oil is generally applied to a petroleum that has an American Petroleum Institute (API) gravity of less than 20° (Chapter 2) and usually, but not always, has a sulfur content higher than 2% by weight. Furthermore, in contrast to conventional crude oils, heavy oils are darker in color and may even be black. They usually contain substantial portions of asphaltenes and resins which are the nonvolatile fractions of petroleums and which contain the highest proportions of heteroatoms (i.e., sulfur, nitrogen, and oxygen). Briefly, the asphaltene fraction of petroleums, heavy oils, and bitumens is that portion of the material which is precipitated when a large excess (40 volumes) of a low-boiling liquid hydrocarbon (e.g., pentane) is added to (1 volume) of the petroleum or heavy oil. The asphaltenes are dark brown to black amorphous solids which do not melt prior to decomposition and are soluble in benzene or aromatic naphthas. On the other hand, resins are isolated from the deasphalted oil by adsorbtion on to an earth (fuller's earth, attapulgus clay, etc.). The light oil fraction is removed from the earth by washing with a hydrocarbon liquid, but the resins can only be removed from the earth by washing with a more polar solvent such as benzene, pyridine, chloroform, etc. The resins are dark red to black semisolid materials which also contain substantial proportions of heteroatoms but, unlike the asphaltenes, are soluble in low-boiling liquid hydrocarbons.

A crude oil residuum (sometimes referred to as the asphaltic fraction) is broadly defined as the residue obtained from the oil after a nondestructive distillation has removed all of the volatile materials. The temperature of the distillation is usually maintained below 350°C (660°F) as it is generally recognized that the rate of thermal decomposition of petroleum constituents is substantial above 350°C. It should be noted here that in certain specific tests for petroleum products (American Society for Testing and Materials test D-86) a temperature up to 370°C (700°F), or until a decomposition point is observed, is recommended.

Residua are black, viscous materials and are thus obtained by atmospheric or vacuum distillation of a crude oil; they may be liquid at room temperature (generally atmospheric residua) or almost solid (generally vacuum residua) depending upon the nature of the crude oil. When a residuum is obtained from a crude oil but thermal decomposition has commenced, it may be more correct to refer to this product as pitch. The chemical composition of a residuum from an asphaltic crude oil is complex; physical methods of fractionation usually indicate high proportions of asphaltenes and resins — even in amounts up to 50% (or higher) of the residuum. In addition, the presence of ash-forming metallic constituents, including such organometallic compounds as those of vanadium and nickel, are also a distinguishing feature of residua and the heavier asphaltic oils

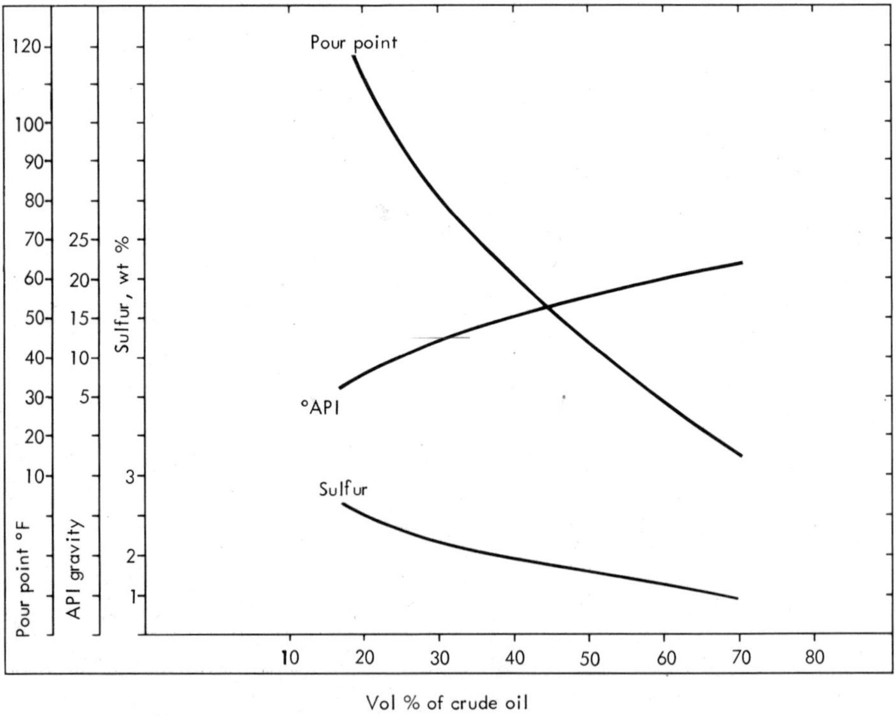

FIG. 1-1 Relation of physical properties to the yield of residuum from Tia Juana (Venezuela) light crude oil.

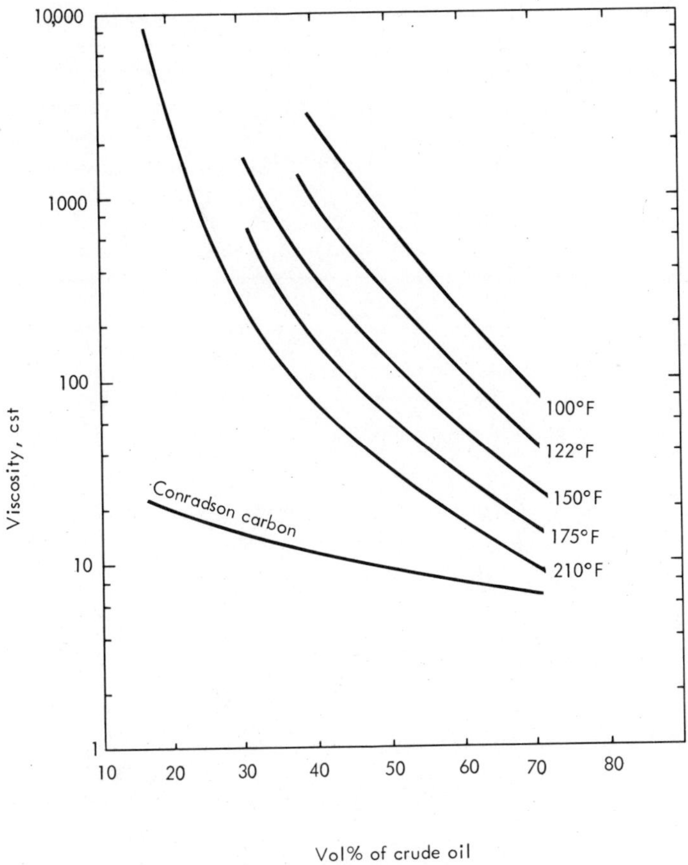

FIG. 1-2 Relation of physical properties to the yield of residuum from Tia Juana (Venezuela) light crude oil.

(Table 1-1). Furthermore, the deeper the "cut" into the petroleum, the greater the concentration of sulfur and metals in the residuum and the greater the deterioration in physical properties (Table 1-2; Figs. 1-1, 1-2, and 1-3).

For a variety of reasons, it is often necessary to define petroleums in terms of their physical composition. Furthermore, the physical composition of heavy oils and residua vary markedly with the method of isolation or separation, thereby leading to further complications in the choice of suitable processing schemes for these feedstocks. Thus, if a petroleum or heavy oil can be defined (on a relative basis) in terms of three general

TABLE 1-1 Miscellaneous Properties of Residua and Heavy Oils

Crude oil origin . . .	Kuwait[a]	Kuwait		Venezuela (east)
Residuum type	Vacuum Gas Oil	Atmospheric	Vacuum	Atmospheric
Fraction of crude, vol %	–	42	21	74
Gravity, °API	22.4	13.9	5.5	9.6
Viscosity SUS, 210° F	–	–	–	–
SFS, 122° F	–	553	500,000	27,400
SFS, 210° F	–	–	–	–
cSt, 100° F	–	–	–	–
cSt, 210° F	–	55	1900	–
Pour point, °F	–	65	–	95
Sulfur, wt %	2.97	4.4	5.45	2.6
Nitrogen, wt %	0.12	0.26	0.39	0.61
Metals, ppm				
Nickel	0.2	14	32	94
Vanadium	0.04	50	102	218
Asphaltenes, wt %				
Pentane insolubles	0	–	11.1	–
Hexane insolubles	0	–	–	–
Heptane insolubles	0	2.4	7.1	9
Resins, wt %	0	–	39.4	–
Carbon residue, wt %				
Ramsbottom	<0.1	9.8	–	14.5
Conradson	0.09	12.2	23.1	–

[a] Included for comparison.

Residua					
Burzurgan	Boscan	Khafji		California	Cabimas
–	–	Atmospheric	Vacuum	Vacuum	Vacuum
52	78	–	–	23	34
3.1	5.0	14.4	6.5	43	6.8
–	–	–	–	–	–
–	–	–	–	–	–
–	–	429	–	–	–
–	–	–	–	–	–
3355	5250	–	–	15,000	7800
–	–	–	–	–	162
6.2	5.9	4.1	5.3	2.3	3.26
0.45	0.79	–	–	0.98	0.62
76	133	37	53	120	76
233	1264	89	178	180	614
–	–	–	12.0	19.0	12.9
–	–	–	–	–	–
18.4	15.3	–	–	–	10.5
–	–	–	–	–	–
–	–	–	–	–	–
22.5	18.0	–	21.4	24.0	18.7

TABLE 1-1 (Continued)

Crude oil origin ...	Arabian light	Safaniya	Heavy Louisiana
Residuum type	Atmospheric	Atmospheric	Vacuum
Fraction of crude, vol %	–	–	13.1
Gravity, °API	16.7	14.5	11.3
Viscosity SUS, 210°F	–	–	–
SFS, 122°F	–	–	–
SFS, 210°F	–	–	–
cSt, 100°F	–	–	–
cSt, 210°F	27	–	700
Pour point, °F	–	–	–
Sulfur, wt %	3.00	3.94	0.93
Nitrogen, wt %	–	0.24	0.38
Metals, ppm			
Nickel	11	20	20
Vanadium	28	78	5
Asphaltenes, wt %			
Pentane insolubles	–	–	6.5
Hexane insolubles	–	–	–
Heptane insolubles	2.0	–	–
Resins, wt %	–	–	–
Carbon residue, wt %			
Ramsbottom	8	11.8	12.0
Conradson	–	–	–

oil and bitumens

Saudi Arabia	Alaska (North Slope)		Boscan
Vacuum	Atmospheric	Vacuum	–
20	58	22	100
5.0	15.2	8.2	10.3
–	1281	–	–
–	–	–	–
–	–	–	–
–	–	–	20,000
2700	42	1950	–
–	75	–	37
5.2	1.6	2.2	5.6
0.30	0.36	0.63	–
28	18	47	117
75	30	82	1220
15.0	4.3	8.0	12.6
–	–	–	11.4
–	31.5	–	–
–	–	–	24.1
20.0	8.4	17.3	14.0
–	–	–	–

TABLE 1-1 (Continued)

	Heavy		
Crude oil origin . . .	Tar Sand Triangle	P.R. Spring	N.W. Asphalt Ridge
Residuum type	–	–	–
Fraction of crude, vol %	100	100	100
Gravity, °API	11.1	10.3	14.4
Viscosity SUS, 210° F	–	–	–
SFS, 122° F	–	–	–
SFS, 210° F	–	–	–
cSt, 100° F	7000[b]	200,000[b]	15,000[b]
cSt, 210° F	–	–	–
Pour point, ° F	–	–	–
Sulfur, wt %	4.38	0.75	0.59
Nitrogen, wt %	0.46	1.00	1.02
Metals, ppm			
Nickel	53	98	120
Vanadium	108	25	25
Asphaltenes, wt %			
Pentane insolubles	26.0	16.0	6.3
Hexane insolubles	–	–	–
Heptane insolubles	–	–	–
Resins, wt %	–	–	–
Carbon residue, wt %			
Ramsbottom	–	–	–
Conradson	21.6	12.5	3.5

[b] Estimated.

oil and bitumens

Athabasca	Lloydminster	Cold Lake	Qayarah
–	–	–	–
100	100	100	100
5.9	14.5	10.0	15.3
513	260	–	–
–	294	–	–
820	–	–	–
–	–	–	–
–	–	79	–
50	38	–	–
4.9	4.3	4.4	8.4
0.41	–	0.39	0.7
86	40	62	60
167	100	164	130
17.0	12.9	15.9	20.4
13.5	–	–	14.3
11.4	–	10.8	13.5
34.0	38.4	31.2	36.1
14.9	–	–	–
18.5	9.1	13.6	15.6

TABLE 1-2 Characteristics of Various Residua Produced from Tia Juana (Venezuela) Light Crude Oil

Boiling range									
°F	Whole crude	>430	>565	>650	>700	>750	>850	>950	>1050
°C	Whole crude	>220	>295	>345	>370	>400	>455	>510	>565
Yield on crude, vol %	100.0	70.2	57.4	48.9	44.4	39.7	31.2	23.8	17.9
Gravity, °API	31.6	22.5	19.4	17.3	16.3	15.1	12.6	9.9	7.1
Specific gravity	0.8676	0.9188	0.9377	0.9509	0.9574	0.9652	0.9820	1.007	1.0209
Sulfur, wt %	1.08	1.42	1.64	1.78	1.84	1.93	2.12	2.35	2.59
Carbon residue (Conradson), wt %	–	6.8	8.1	9.3	10.2	11.2	13.8	17.2	21.6
Nitrogen, wt %	–	–	–	0.33	0.36	0.39	0.45	0.52	0.60
Pour point, °F	-5	15	30	45	50	60	75	95	120

Viscosity:									
Kinematic, cSt	10.2	83.0	315	890	1590	3100	—	—	—
@ 100° F	10.2	83.0	315	890	1590	3100	—	—	—
@ 210° F	—	9.6	19.6	35.0	50.0	77.0	220	1010	7959
Furol (SFS) sec									
@ 122° F	—	—	70.6	172	292	528	—	—	—
@ 210° F	—	—	—	—	25.2	37.6	106	484	3760
Universal (SUS) sec @ 210° F	—	57.8	96.8	165	234	359	1025	—	—
Metals:									
Vanadium, ppm	—	—	—	185	—	—	—	—	450
Nickel, ppm	—	—	—	25	—	—	—	—	64
Iron, ppm	—	—	—	28	—	—	—	—	48

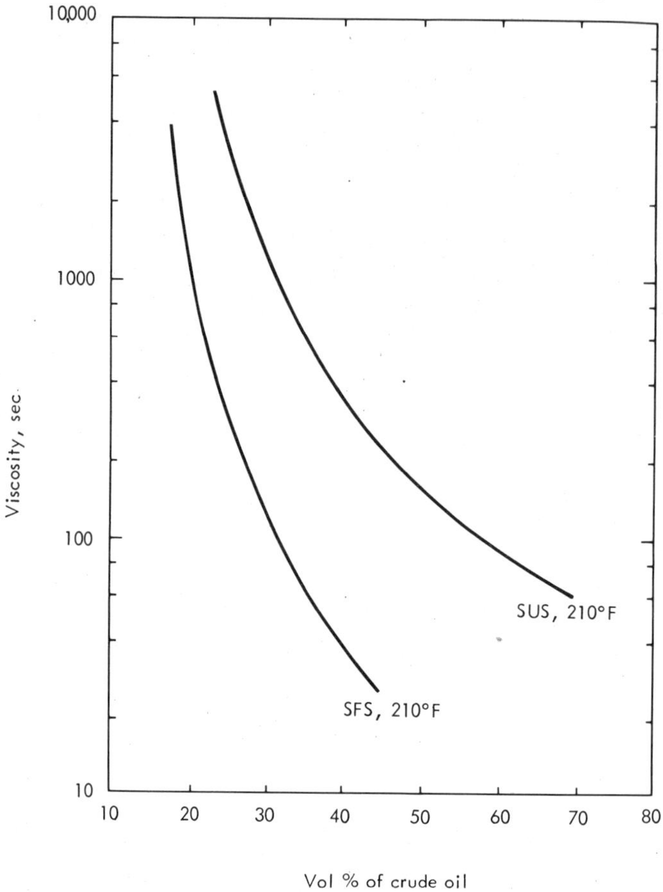

FIG. 1-3 Relation of physical properties to the yield of residuum from Tia Juana (Venezuela) light crude oil.

fractions—asphaltenes, resins, and oils (Fig. 1-4)—residua can also be defined in terms of these fractions but usually with substantial variations in the relative proportions of these fractions. Furthermore, there will also be variations of the compound types within these three fractions, especially in the composition of the oils fraction where the more volatile constituents will be removed by distillation if the residuum is, in fact, prepared by atmospheric or vacuum distillation of the petroleum. In addition, production of the residuum by a solvent technique will yield a product that is somewhat different from the distillation residuum insofar as the proportion of the oils fraction will be reduced (Fig. 1-5).

The two major characteristics of residua and heavy oils which distinguish them from distillates are that residua and heavy oils contain (1) asphaltenes and other high-molecular-weight, highly aromatic, structures and (2) ash-forming constituents, including organometallic compounds, with nickel and vanadium being the predominant metals. Asphaltenes have a marked effect on refining and result in the deposition of high yields of coke during the various thermal processes and, insofar as coke will act as an insulator, increasing layers of coke in a reactor will demand an increasing heat input, leading to substantial localized, or even general, overheating and, consequently, even more coke deposition. In the present context, the deposition of coke on a desulphurization catalyst will seriously affect catalyst activity with a marked decrease in the rate of desulphurization. In fact, it has been noted that even with a deasphalted feedstock, i.e., a heavy feedstock from which the asphaltenes have previously been removed, the accumulation of carbonaceous deposits on the catalyst is still substantial. It has been suggested that this deposition of carbonaceous material is due to the condensation reactions that are an integral part of any thermal (even hydrocracking) process in which heavy feedstocks are involved. It appears that the large molecules originally present in the feedstock (or formed

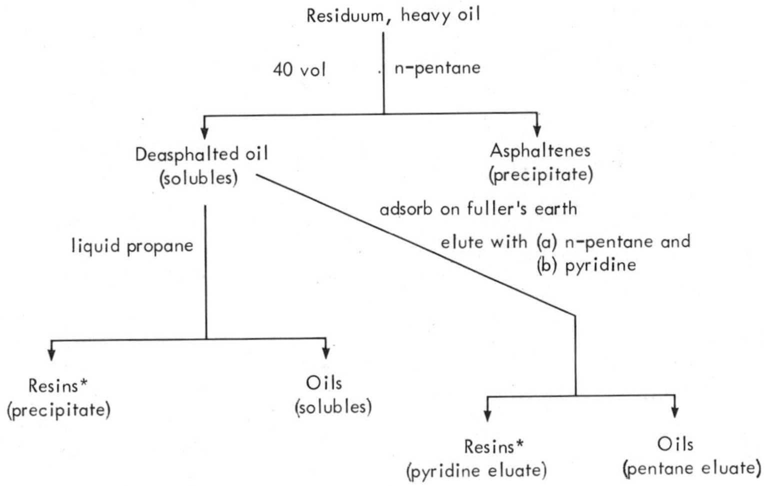

*The resins isolated by one technique may well vary somewhat from the resins isolated by the alternate technique; the same applies to the oils fractions.

FIG. 1-4 Schematic representation of the fractionation of heavy oils and residua into the three broad fractions—asphaltenes, resins, and oils.

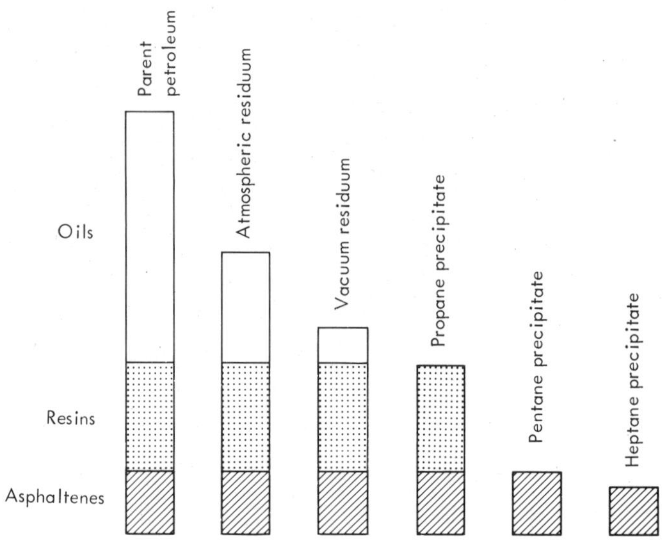

FIG. 1-5 Simplified relationship between a parent petroleum, the atmospheric and vacuum residua, and the various solvent fractions; a heavy oil may be approximated to the composition of the atmospheric or vacuum residua. [Reprinted with permission from L. W. Corbett and U. Petrossi, Ind. Eng. Chem., Prod. Res. Dev., 1978, 17(4):346. Copyright by the American Chemical Society.]

during the process) are not sufficiently mobile (or are too strongly adsorbed by the catalyst) to be saturated by the hydrogenation components and, hence, continue to condense and eventually degrade to coke. These deposits deactivate the catalyst sites and eventually interfere with the hydrodesulfurization process. Thus, the deposition of coke and, hence, the rate of catalyst deactivation, is subject to variations in the asphaltene (and resins) content of the feedstock as well as the adsorptive properties of the catalyst for the heavier molecules.

The second characteristic of residua and heavy oils pertinent to hydrodesulfurization is the presence of organometallic compounds (of which nickel and vanadium are the principal constituents) that are present to varying degrees in all residua and in the majority of heavy oils. Deposition of these metals in any form on to the catalyst leads to catalyst deactivation although the exact mechanism of deactivation is still subject to speculation. However, nickel does tend to be deposited throughout the catalyst whereas vanadium is more usually concentrated in the outer layers of the catalyst. In either case, catalyst deactivation is certain whether it be by physical blockage of the pores or destruction of reactive sites.

1-2 THE CHEMICAL COMPOSITION OF HEAVY OILS AND RESIDUA

The chemical composition of heavy oils and residua is, in spite of the large volume of work performed in this area, largely speculative. Indeed, the "simpler" petroleums are extremely complex mixtures of organic compounds. In fact, the composition of petroleum can vary with the location and age of the field in addition to any variations that occur with the depth of the individual well. Two adjacent wells are more than likely to produce petroleum with very different characteristics.

On a molecular basis, petroleum contains hydrocarbons as well as the organic compounds of sulfur, nitrogen, and oxygen; metallic constituents may also be present, but only to a minor extent. While the hydrocarbon content of petroleum may be as high as 97% (as, for example, in the lighter paraffinic crude oils), it is, nevertheless, the nonhydrocarbon (i.e., sulfur, oxygen, and nitrogen) constituents that play a large part in determining the nature and, hence, the processability of the crude oil.

The elemental composition (ultimate analysis) of petroleum, no matter what the origin of the particular petroleum, varies only slightly over very narrow limits:

Carbon	83.0 -87.0%
Hydrogen	10.0 -14.0
Sulfur	0.05- 6.0
Nitrogen	0.1 - 2.0
Oxygen	0.05- 1.5

Thus, it is not possible to classify petroleum in a manner similar to the classification of coal on the basis of carbon content and the range of elemental compositions for crude oils includes those lighter, more mobile, paraffinic materials at one extreme to the heavier asphaltic crude oils and bitumens at the other extreme.

Nevertheless, the high proportion of carbon and hydrogen in petroleum indicate that hydrocarbons are the major constituents and may actually account for more than 75% of the constituents of many petroleums. The inclusion of organic compounds of sulfur, nitrogen, and oxygen only serves to present crude oils as even more complex mixtures, and the appearance of appreciable amounts of these nonhydrocarbon compounds causes some concern in the refining of crude oils. Even though the concentration of nonhydrocarbon constituents (i.e., those organic compounds containing one or more sulfur, oxygen, or nitrogen atom) in certain fractions may be quite small, they do, however, tend to concentrate in the higher-boiling fractions of petroleum. Indeed, their influence on the processability of the petroleum is important irrespective of their molecular size and the fraction in which they occur.

Thus, the presence of traces of nonhydrocarbon compounds can impart objectionable characteristics to finished products leading to discoloration and/or lack of stability during storage. On the other hand, catalyst poisoning and corrosion will be the most noticeable effects during refining sequences when these compounds are present. It is, therefore, not surprising that considerable attention must be given to the nonhydrocarbon constituents of petroleum as the trend in the refining industry, of late, has been to process the heavier crude oils as well as residua which contain substantial proportions of these nonhydrocarbon materials.

A. Sulfur Compounds

Sulfur compounds are perhaps the most important nonhydrocarbon constituents of petroleum. During the refining sequences involved to convert crude oils to salable products, a great number of the sulfur compounds that occur in any particular petroleum are concentrated in the residua and other heavy fractions. Indeed, the relationship of sulfur content to asphaltics (asphaltenes plus resins) content of crude oils is well documented (Fig. 1-6) and it is not surprising that some attention has been given to the processing of these heavy, high-sulfur, fractions of crude oil.

The nature of the sulfur compounds in petroleum is receiving considerable attention, and some success has been recorded in attempts to identify individual compounds from a variety of petroleums. There are many varieties of sulfur compounds (Table 1-3), but the prevailing conditions during the formation, maturation, and any in situ alteration may dictate the only preferred types exist in any particular crude oil. Nevertheless, sulfur compounds of one type or another are generally present in all crude oils.

The relative importance attached to sulfur compounds in petroleum may, at first, seem unwarranted, but the presence of sulfur compounds in any crude oil can only result in harmful effects. For example, the presence of sulfur compounds in finished petroleum products such as gasoline will cause corrosion of engine parts, especially under winter conditions when water containing sulfur dioxide (from the internal combustion) may collect in the crankcase. On the other hand, mercaptans cause the corrosion of copper and brass in the presence of air and also have an adverse effect on the color stability of gasolines and other liquid fuels.

Free sulfur is also corrosive as are sulfides, disulfides, and thiophenes which are detrimental to the octane number response to tetraethyllead. Sulfur compounds are also detrimental to diesel fuels by increasing engine wear as well as contributing to the formation of engine deposits.

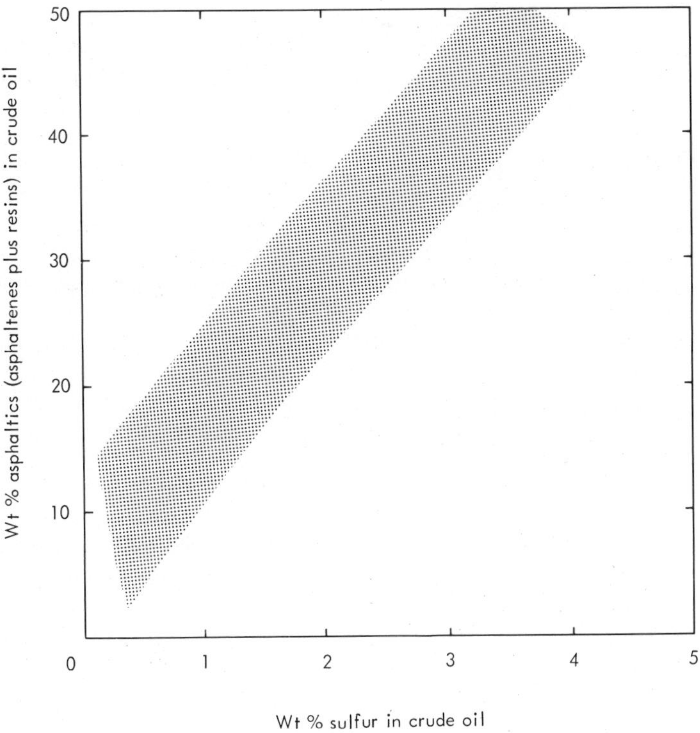

FIG. 1-6 Relation of asphaltics content to sulfur content of crude oils.

Similarly, a high content of sulfur compounds in lubricating oils lowers resistance of the oil to oxidation and increases deposition of solid material. A high sulfur content can sometimes be tolerated in industrial fuel oils, but current environmental regulations dictate that there now be little, if any, sulfur compounds present in the majority of fuels.

The distribution of sulfur compounds in crude oils has been studied extensively since the 1890s and it has become possible to note various generalities. For example, the proportion of sulfur will increase with the boiling point of the crude oil fraction (Table 1-4), but if the distillation is allowed to proceed at too high a temperature, thermal decomposition of the high-molecular-weight sulfur compounds will ensue and, hence, the middle fractions will contain more sulfur than the higher-boiling fractions. The distribution of the various types of sulfur compounds varies markedly among crude oils of diverse origin and it is difficult to assign specific trends to the occurrence of compound types within the different crude oils

TABLE 1-3 Nomenclature and Types of Organic Sulfur Compounds

H–S–H	Hydrogen sulfide
Mercaptans:	
$H{-}S{-}CH_3$	Methyl mercaptan
$H{-}S{-}C_6H_5$	Phenyl mercaptan
Sulfides:	
$CH_3{-}S{-}CH_3$	Dimethyl sulfide
$C_4H_9{-}S{-}C_4H_9$	Di-n-butyl sulfide
Disulfides:	
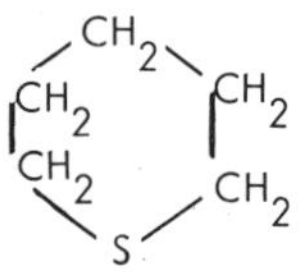	Thiacytohexane (pentamethylene sulfide)
Alkyl sulfates:	
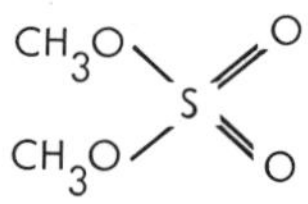	Dimethyl sulfate
Sulfonic acids:	
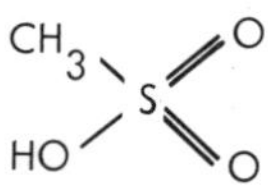	Methyl sulfonic acid
Sulfoxides:	
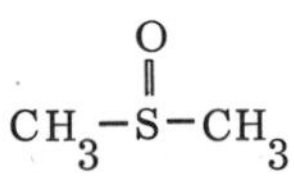	dimethyl sulfoxide
Sulfones:	
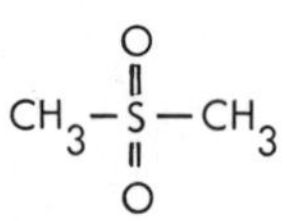	Dimethyl sulfone

TABLE 1-3 (Continued)

Thiophenes:

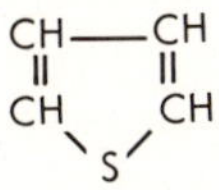	Thiophene
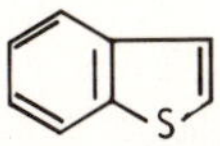	Benzothiophene

Source: Reproduced with permission from J. G. Speight, The Chemistry and Technology of Petroleum, 1980, p. 66, courtesy Marcel Dekker, Inc.

other than an increase in boiling point of fractions from a particular crude oil is accompanied by an increase in sulfur content.

Elemental sulfur also occurs in crude oils and participates in thermal reactions at relatively low temperatures. Thus, if elemental sulfur is present in an oil, a reaction, with the evolution of hydrogen sulfide, begins at about 150° C (300° F) and is very rapid at 220° C (430° F), but organically bound sulfur is not released as hydrogen sulfide until higher temperatures are attained. It should be noted at this point that hydrogen sulfide is, in fact, a common constituent of many crude oils but the origin of this sulfur-containing gas is subject to much speculation. Nevertheless, crude oils containing more than 1% sulfur are often accompanied by a gas having substantial proportions of hydrogen sulfide.

The identification of sulfur-containing constituents of crude oils has been the subject of many investigations with the result that many of the sulfur compound types that occur in crude oils are now known. These vary from the simple thiols:

CH_3CH_2SH

Ethanethiol

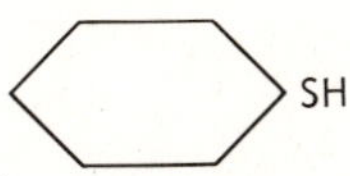

Cyclohexylthiol

to the simple sulfides:

$CH_2CH_2SCH_2CH_3$

Diethyl sulfide
(3-Thiapentane)

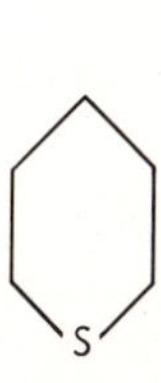

Thiacyclohexane

TABLE 1-4 Sulfur Distribution in Petroleum Fractions

Crude	Sulfur content of crude oil (wt %)	Sulphur in Product (% of original sulfur in the oil)			
		Naphtha (Gasoline)	Kerosene	Gas Oil	Residuum
Far East	0.15	0.3	3.6	38.6	57.5
East Texas	0.36	0.9	1.3	15.4	82.4
East Venezuela	0.55	0.5	1.7	15.5	82.3
Iran	1.4	1.1	1.5	12.6	84.8
West Texas	2.0	1.8	4.2	14.8	79.2
West Venezuela	2.2	0.05	0.55	6.6	92.8
Kuwait	2.45	0.1	0.8	9.5	89.6

and the polycyclic sulfides, such as compounds with two condensed rings:

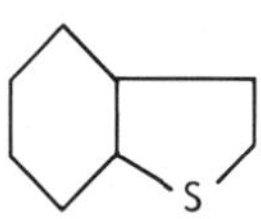

Thiahydrindane

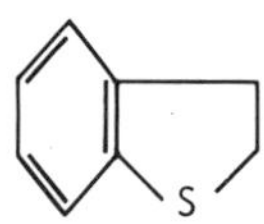

1-Thiaindan

which have all been isolated from petroleum. Various thiophenes, e.g.:

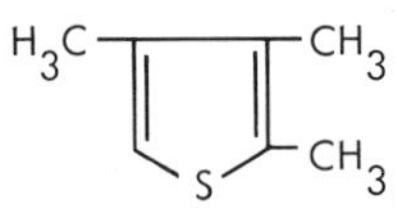

2, 3, 4-Trimethylthiophene

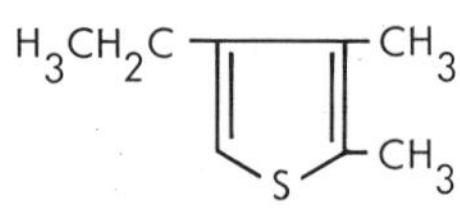

2, 3-Dimethyl-4-ethylthiophene

have also been isolated from a variety of crude oils while benzothiophene derivatives:

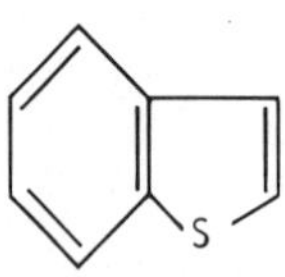

Benzothiophene

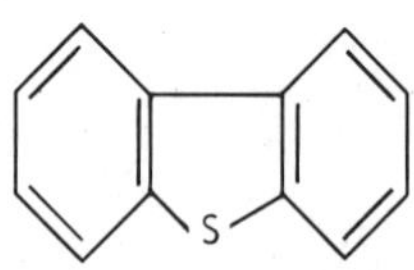

Dibenzothiophene

are usually present in the higher-boiling petroleum fractions. On the other hand, disulfides are not regarded as being true constituents of crude oil but are generally formed by oxidation of thiols during processing:

$$2R-SH \xrightarrow{[O]} R-S-S-R + H_2O$$

Although sulfur is the most important heteroatom (i.e., nonhydrocarbon) present in petroleum with respect to the current context, there are other nonhydrocarbon atoms that can exert a substantial influence not only on the nature and properties of the products but also on the nature and efficiency of the process. Such atoms are nitrogen, oxygen, and metals, and because of their influence on the process, some discussion of each is warranted here.

B. Nitrogen Compounds

Nitrogen in petroleum may be classed arbitrarily as "basic" and "nonbasic." The basic nitrogen compounds (Table 1-5), which are composed mainly of pyridine homologs and occur throughout the boiling ranges, have a decided tendency to exist in the higher-boiling fractions and residua (Fig. 1-7). The nonbasic nitrogen compounds, which are usually of the pyrrole, indole, and carbazole types, also occur in the higher-boiling fractions and residua.

In general, the nitrogen content of petroleums is low and falls within the range 0.1 to 0.9%, although early work indicates that some crudes may contain up to 2% nitrogen. However, crude oils with no detectable nitrogen or even trace amounts are not uncommon but, in general, the more asphaltic the oil, the higher its nitrogen content. Insofar as an approximate correlation exists between the sulfur content and API gravity of crude oils (Chapter 2), there also exists a correlation between nitrogen content and the API gravity of crude oil (Fig. 1-8). It also follows that there is an approximate correlation between the nitrogen content and the carbon residue—the higher the carbon residue, the higher the nitrogen content. The presence of nitrogen in petroleum is of much greater significance in refinery operations than might be expected from the small amounts present. Nitrogen compounds can be responsible for the poisoning of cracking catalysts, and they also contribute to gum formation in such products as domestic fuel oil. The trend in recent years toward cutting deeper into the crude to obtain stocks for catalytic cracking has accentuated the harmful effects of the nitrogen compounds which are concentrated largely in the higher-boiling portions.

Basic nitrogen compounds having relatively low molecular weight can be extracted with dilute mineral acids, while equally strong bases of higher molecular weight remain unextracted because of unfavorable partition between the oil and aqueous phases. A method has been developed in which

TABLE 1-5 Nomenclature and Types of the Common Organic Nitrogen Compounds

Structure	Name
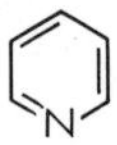	Pyridine
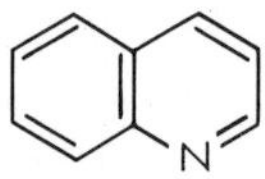	Quinoline
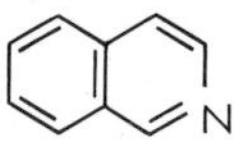	Isoquinoline
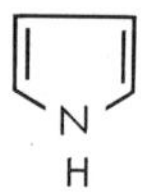	Pyrrole
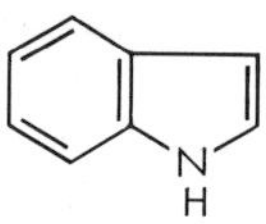	Indole
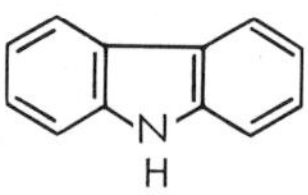	Carbazole

Source: Reproduced with permission from J. G. Speight, *The Chemistry and Technology of Petroleum*, 1980, p. 74, courtesy Marcel Dekker, Inc.

the nitrogen compounds are classified as basic or nonbasic, depending on whether they can be titrated with perchloric acid in a 50-50 solution of glacial acetic acid and benzene. Application of this method showed that the ratio of "basic" to total nitrogen is approximately constant (0.30 ± 0.05) irrespective of the source of the crude. Indeed, the ratio of basic to total nitrogen was found to be approximately constant throughout the entire range of distillate and residual fractions. Nitrogen compounds extractable with dilute mineral acids from petroleum distillates were found to consist of pyridines, quinolines, and isoquinolines carrying alkyl substituents, as well as a few pyridines in which the substituent was a cyclopentyl or cyclohexyl group. The compounds which cannot be extracted with dilute mineral acids contain the greater part of the nitrogen in petroleum and are generally of the carbazole, indole, and pyrrole type.

Porphyrins (nitrogen-metal complexes) are also constituents of petroleum and usually occur in the nonbasic portion of the nitrogen-containing concentrate. Pyrrole, the chief constituent of the porphyrin molecule, is marked by high stability due to its aromatic character. As a result of aromatization arising from the conjugated bond system, pyrrole is not strongly basic even though it is a secondary amine. Pyrrole, like other heterocyclic molecules, differs from homocyclic aromatic

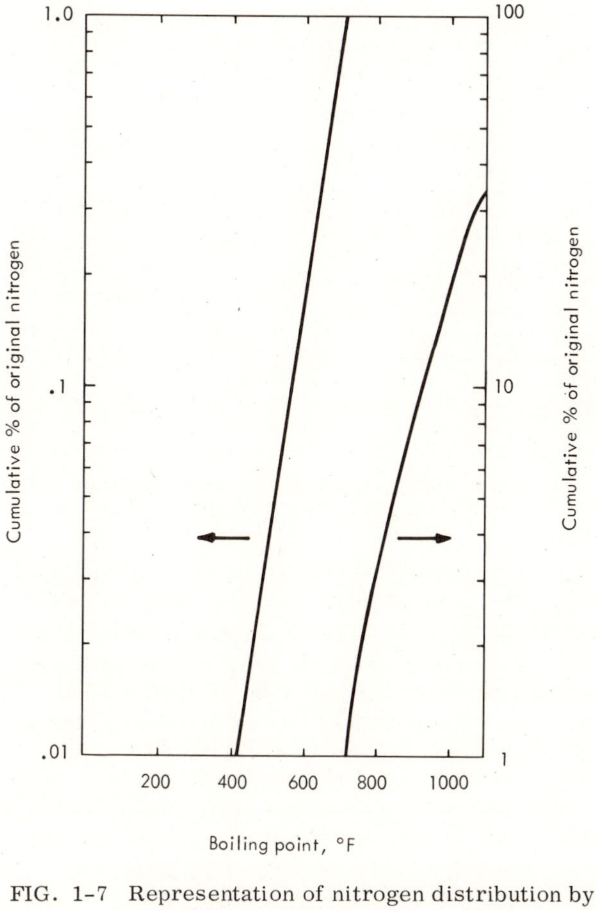

FIG. 1-7 Representation of nitrogen distribution by boiling point. (Reproduced with permission from J. H. Gary and G. H. Handwerk, Petroleum Refining—Technology and Economics, 1975, p. 34, courtesy of Marcel Dekker, Inc.)

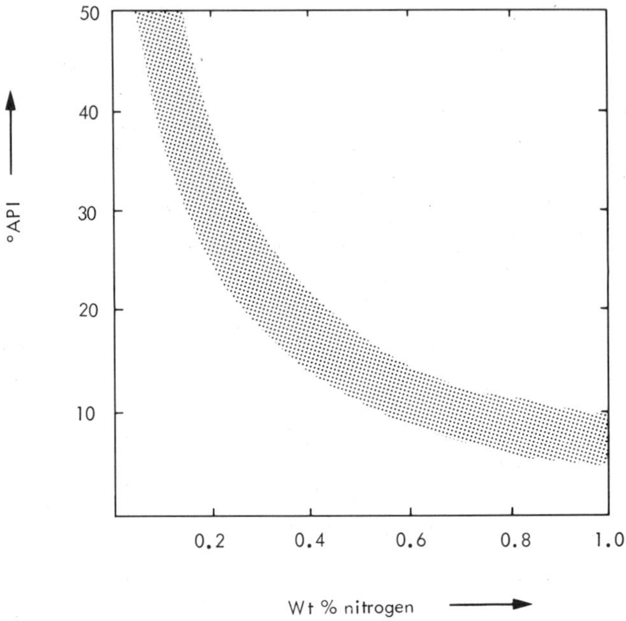

FIG. 1-8 The relationship of nitrogen content of crude oils to API gravity. (Reproduced with permission from J. G. Speight, The Chemistry and Technology of Petroleum, 1980, courtesy of Marcel Dekker, Inc.)

compounds in that it is quite reactive at the position alpha to the nitrogen atom and tends to form dimers, trimers, and higher condensation products in which the fundamental pyrrole structure is preserved to a marked degree.

The simplest porphyrin is porphine and consists of four pyrrole molecules joined by methine (—CH=) bridges (Fig. 1-9). The methine bridges establish conjugated linkages between the component pyrrole nuclei forming a more extended resonance system. Although the resulting structure retains much of the inherent character of the pyrrole components, the larger conjugated system gives increased aromatic character to the porphine molecule. The reactivity of the pyrrole nuclei in porphine is greatly reduced, partially because of the increased aromatic character but primarily because the reactive alpha positions are occupied. Two closely related classes of porphyrin-type compounds are the dihydroporphyrins and the tetrahydroporphyrins. The dihydroporphyrins, commonly known as chlorins, include two "extra" hydrogen atoms on one of the pyrrole rings and the tetrahydroporphyrins carry four "extra" hydrogen atoms.

The most common metal complexes of porphyrins are those formed by replacement of the two "nitrogen" hydrogens with a relatively small cation. Indeed, metal complexes of porphyrins are widespread in petroleum. Their concentrations are quite low (Table 1-6) and in some oils porphyrins do not appear to be present.

The presence of vanadium and nickel in crude oils, especially as metal porphyrin complexes, has focused much attention in the petroleum refining industry on the occurrence of these metals in feedstocks. Only a part of the total nickel and vanadium in crude oil is recognized to occur in porphyrin structures (Table 1-6). In general, it is assumed that about 10% of the total metal in a crude oil is accommodated as porphyrin complexes although as much as 40% of the vanadium and nickel may be present as metal porphyrin complexes in petroleum.

C. Oxygen Compounds

Oxygen in organic compounds can occur in a variety of forms (Table 1-7) and it is not surprising that the more common of the oxygen-containing compounds occur in petroleum. The total oxygen content of petroleums is usually less than 2%, although larger amounts have been reported but in cases where the oxygen content is phenomenally high it may be that the oil has suffered prolonged exposure to the atmosphere either during or after production. However, the oxygen content of petroleum does increase with the boiling point of the fractions examined; in fact, the nonvolatile residua may have oxygen contents up to 8% by weight. Though these high-molecular-weight compounds contain most of the oxygen in petroleum, little is known concerning their structure, but those of lower molecular weight have been investigated with considerably more success and have been shown to contain carboxylic acids and phenols.

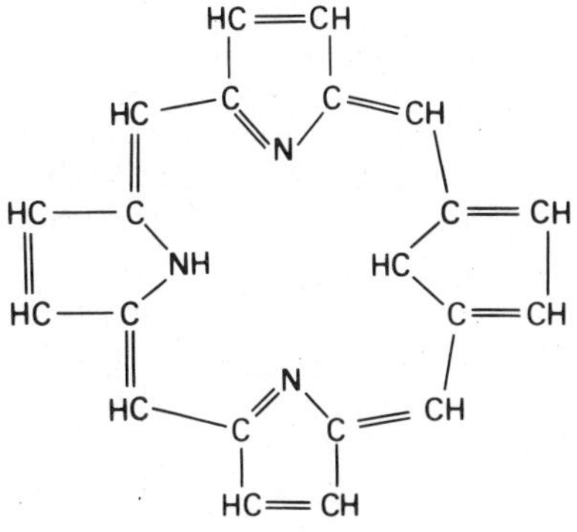

FIG. 1-9 The structure of porphine—the basic structural unit of porphyrins.

TABLE 1-6 Porphyrin Concentrations in Petroleum

Source	Porphyrins (ppm)	Vanadium and Nickel Complexed %	V (ppm)	Ni (ppm)
Bachaquero, Venezuela	380	8.5	430	53
Boscan, Venezuela	1680	18.7	900	66
Coleville, Canada	110	9.6	94	32
Lagunillas, Venezuela	170	5.1	317	41
North Belridge, Calif.	390	44.6	23	83
Rhodes, Kans.	135	8.8	133	33
Santa Maria, Calif.	300	8.1	280	130
Tatums, Okla.	165	8.4	148	71
Tia Juana, Venezuela	60	3.2	187	24
West Texas, Texas	12	4.7	23	5
Wilmington, Calif.	200	21.6	46	60

The presence of acid substances in petroleum first appears to have been reported in 1874 and it was established nine years later that these substances contained carboxyl groups and were carboxylic acids. These were termed "naphthenic acids." Although alicyclic (naphthenic) acids appear to be the more prevalent, it is now well known that aliphatic acids are also present (Table 1-8). In addition to the carboxylic acids, alkaline extracts from petroleum contain phenols.

It has generally been concluded that the carboxylic acids in petroleum with less than eight carbon atoms per molecule are almost entirely aliphatic in nature; monocyclic acids begin at C_6 and predominate above C_{14}. This indicates that the structures of the carboxylic acids correspond with those of the hydrocarbons with which they are associated in the crude oil; that is, in the range where paraffins are the prevailing type of hydrocarbon, the aliphatic acids may be expected to predominate; similarly, in the ranges where the monocycloparaffins and dicycloparaffins prevail, one may expect to find principally monocyclic and dicyclic acids, respectively.

In addition to the carboxylic acids and phenolic compounds, the presence of ketones, esters, ethers, and anhydrides has been claimed for a variety

TABLE 1-7 Nomenclature and Types of the Common Organic Oxygen Compounds

Alcohols:	
CH_3-O-H	Methyl alcohol
C_6H_5-O-H	Phenyl alcohol (phenol)
Ethers:	
CH_3-O-CH_3	Dimethyl ether
$C_6H_5-O-C_6H_5$	Diphenyl ether
Cyclic ethers:	
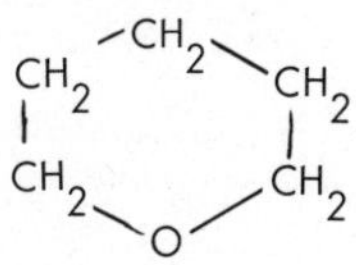	Tetrahydropyran (pentamethylene oxide)
Carboxylic acids:	
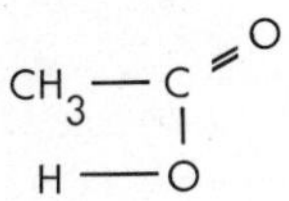	Acetic acid
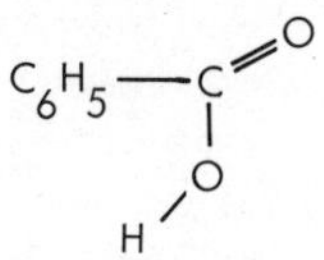	Benzoic acid
Carboxylic acid anhydrides:	
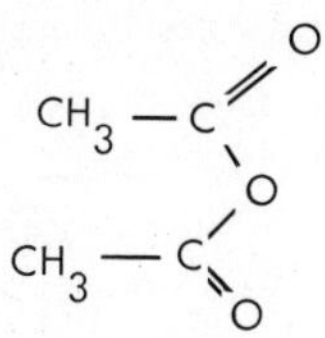	Acetic anhydride
Carboxylic acid esters:	
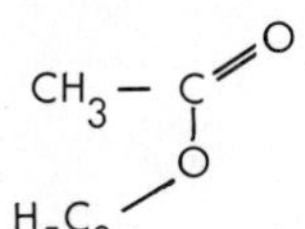	Ethyl acetate

TABLE 1-7 (Continued)

Ketones:	
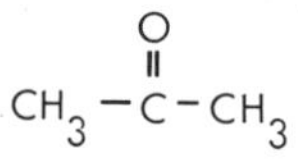	Dimethyl ketone (acetone)
Furans:	
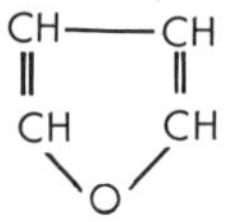	Furan
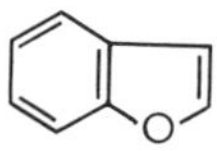	Benzofuran

Source: Reproduced with permission from J. G. Speight, The Chemistry and Technology of Petroleum, 1980, p. 70, courtesy Marcel Dekker, Inc.

of petroleums. However, the precise identification of these compounds is difficult as most of them occur in the higher-molecular-weight nonvolatile residua. They are claimed to be products of the air blowing of the residua, and their existence in virgin petroleum may yet need to be substantiated.

Although comparisons are frequently made between the sulfur and nitrogen contents and physical properties such as the API gravity (Fig. 1-8; see also Chapter 2), it is not the same with the oxygen contents of crude oils. Indeed, even though it may be possible to postulate, and show, that such relationships exist, the ease with which some of the crude oil constituents can react with oxygen (aerial or dissolved) to incorporate oxygen functions into their molecular structure often renders the exercise somewhat futile if meaningful deductions are to be made.

A. Metallic Constituents

The occurrence of metallic constituents in crude oil is of considerably greater interest to the petroleum industry than might be expected from the very small amounts present. Even minute amounts of iron, copper, and particularly nickel and vanadium in the charging stocks for catalytic cracking affect the activity of the catalyst and result in increased gas and coke formation and reduced yields of gasoline. In high-temperature power generators, such as oil-fired gas turbines, the presence of metallic constituents,

TABLE 1-8 Naphthenic Acids in Various Crude Oils

Crude	Type	Fraction	Naphthenic acid (%)
American Crude Oils:			
Pennsylvania	Paraffinic	Kerosene	0.006
Pennsylvania	Paraffinic	Gas oil	0.010
East Texas	Intermediate	Kerosene	0.009
Mid-continent	Intermediate	Kerosene	0.009
California	Naphthenic	Naptha	0.01
		Kerosene	0.06
		Gas oil	0.36
Texas heavy	Naphthenic	Kerosene	0.075
		Gas oil	0.35
Russian Crude Oils:			
Balakhany light	Naphthenic	Whole crude	1.05
		Kerosene	0.5
Balakhany heavy	Asphaltic	Whole crude	1.10
		Kerosene	
Binagady	Asphaltic	Whole crude	0.85
		Kerosene	0.5
Ramain	Intermediate	Whole crude	0.40
		Kerosene	0.20
Surakhani	Intermediate	Whole crude	0.20
		Kerosene	0.20

Source: Reproduced with permission from J. G. Speight, The Chemistry and Technology of Petroleum, 1980, p. 72, courtesy Marcel Dekker, Inc.

particularly vanadium in the fuel, may lead to ash deposits on the turbine rotors, thus reducing clearances and disturbing their balance. More particularly, damage by corrosion may be very severe. The ash resulting from the combustion of fuels containing sodium and especially vanadium reacts with refractory furnace linings to lower their fusion points, and so cause their deterioration.

Thus, the ash residue left after burning of a crude oil is due to the presence of these metallic constituents, part of which occur as inorganic water-soluble salts (mainly chlorides and sulfates of sodium, potassium, magnesium, and calcium) and which occur in the water phase of crude oil emulsions. These are removed in the desalting operations, either by evaporation of the water and subsequent water washing, or by breaking the emulsion, thereby causing the original mineral content of the crude to be substantially reduced. Other metals are present in the form of oil-soluble organometallic compounds either as complexes, metallic soaps, or in the form of colloidal suspensions, and the total ash from desalted crudes is of the order of 0.1 to 100 mg liter^{-1}.

Two groups of elements appear in significant concentrations in the original crude oil, associated with well-defined types of compounds. Zinc, titanium, calcium, and magnesium appear in the form of organometallic soaps with surface-active properties, adsorbed in the water-oil interfaces, and act as emulsion stabilizers. However, vanadium, copper, nickel, and part of the iron found in crude oils seem to be in a different class and are present as oil-soluble compounds. These metals are capable of complexing with pyrrole pigment compounds derived from chlorophyll and hemoglobin are almost certain to have been present in plant and animal source materials. It is easy to surmise that the metals in question are present in such form, ending in the ash content. Evidence for the presence of several other metals in oil-soluble form has been produced and, thus, zinc, titanium, calcium, and magnesium compounds have been identified in addition to vanadium, nickel, iron, and copper (Table 1-9). Examination of the analyses of a number of crudes for iron, nickel, vanadium, and copper (Table 1-10) indicates a relatively high vanadium content which usually exceeds that of nickel, although the reverse can also occur.

Distillation concentrates the metallic constituents in the residua (Table 1-2); some can appear in the higher-boiling distillates but the latter may, in part, be due to entrainment. Nevertheless, there is evidence that a portion of the metallic constituents may occur in the distillates by volatilization of the organometallic compounds present in the petroleum. In fact, as the percentage overhead obtained by vacuum distillation of a reduced crude is increased, the amount of metallic constituents in the overhead oil is also increased. The majority of the vanadium, nickel, iron, and copper in residual stocks may be precipitated along with the asphaltenes by hydrocarbon solvents. Thus, removal of the asphaltenes with n-pentane reduces the vanadium content of the oil by up to 95% with substantial reductions in the amounts of iron and nickel.

1-3 CHEMICAL STRUCTURES IN HEAVY OILS AND RESIDUA

In spite of numerous investigations, the determination of the actual molecular structures that exist in heavy oils and residua has proved to be difficult.

TABLE 1-9 Ranges of Principal Trace Elements Found in Petroleum

Element	Range in Petroleum (ppm)
Cu	0.2 - 12.0
Ca	1.0 - 2.5
Mg	1.0 - 2.5
Ba	0.001- 0.1
Sr	0.001- 0.1
Zn	0.5 - 1.0
Hg	0.03 - 0.1
Ce	0.001- 0.6
B	0.001- 0.1
Al	0.5 - 1.0
Ga	0.001- 0.1
Ti	0.001- 0.4
Zr	0.001- 0.4
Si	0.1 - 5.0
Sn	0.1 - 0.3
Pb	0.001- 0.2
V	5.0 -1500.0
Fe[a]	0.04 - 120.0
Co	0.001- 12.0
Ni	3.0 - 120.0

[a]Part may be due to contamination by iron-containing equipment.
Source: Reproduced with permission from J. G. Speight, The Chemistry and Technology of Petroleum, 1980, p. 77, courtesy Marcel Dekker, Inc.

It is, no doubt, the great complexity of these materials, which are actually isolated by physical phenomena rather than by virtue of their chemical structures, which has hindered the formulation of individual molecular structures. Nevertheless, the various investigations have brought to light some significant facts about the structures that exist in heavy oils and

TABLE 1-10 Metals Content of Various Crude Oils

Source	Metals (ppm)			
	Fe	Ni	V	Cu
East Texas	3.2	1.7	1.2	0.4
West Texas	5.1	4.8	7.9	0.4
Mirando	7.6	1.9	1.4	0.5
Jackson	4.4	1.8	0.9	0.2
Scurry County	3.4	1.0	0.8	0.2
Wilmington	28.0	46.0	41.0	0.6
Santa Maria	17.0	97.0	223.0	0.3
Kettleman	24.0	35.0	34.0	0.4
Ventura	31.0	33.0	49.0	1.1
Tibu-Petrolea	1.6	9.0	60.0	0.9
Kuwait	0.7	6.0	22.5	0.1
Mid-Continent	3.8	4.2	7.9	0.3
Kansas	5.8	5.8	20.8	0.4
Morocco	–	0.8	0.6	0.1
Redwater, Canada	3.4	10.6	4.5	0.1

Source: Reproduced with permission from J. G. Speight, The Chemistry and Technology of Petroleum, 1980, p. 77, courtesy Marcel Dekker, Inc.

residua. There are indications that the higher-molecular-weight components (the asphaltenes and resins) consist of condensed aromatic nuclei which carry alkyl and cylcoalkyl (naphthenic) substituents; heteroatoms (i.e., sulfur, nitrogen, and oxygen) are scattered throughout the hydrocarbon systems. In addition, it appears that with increasing molecular weight, both aromaticity and the proportion of the heteroatoms increase. There is also fragmentary evidence that, with increasing molecular weight, the heteroatoms become incorporated into the more stable heterocyclic systems. This is especially true for sulfur where dibenzothiophene-type systems are believed to predominate in the higher-molecular-weight asphaltics; it is presumed that similar arrangements occur for the nitrogen atoms, while oxygen may also occur in cyclic systems or in various functional groups.

If, however, the residuum is produced by a process where thermal decomposition may have been initiated, there may be traces of thiols in the mix which would, along with any hydrogen sulfide that was generated during the process, impart an objectionable odor to the residuum. The occurrence of thiols in heavy oils that have not been subjected to any thermal treatment is also possible but is generally not considered likely.

BIBLIOGRAPHY

Brooks, B. T., S. S. Kurtz, Jr., C. E. Boord, and L. Schmerling: The Chemistry of Petroleum Hydrocarbons, Reinhold, New York, 1954.

Charbonnier, R. P., R. G. Draper, W. H. Harper, and A. Yates: Analyses and Characteristics of Oil Samples from Alberta, Department of Energy Mines and Resources, Mines Branch, Ottawa, Information Circular IC 232, June 1969.

Considine, D. M., ed.: Energy Technology Handbook, McGraw-Hill, New York, 1977.

Draper, R. G., E. Kowalchuk, and G. Noel: Analyses and Characteristics of Crude Oil Samples Performed between 1969 to 1976, Energy, Mines, and Resources, Canada, Energy Research Laboratories, Report ERP/ERL 77-59 (TR), July 1977.

Druschel, Harry V.: Sulfur compounds in petroleum—known and unknown, Preprints, Amer. Chem. Soc. Div. Petrol. Chem., 1970, 15(2), C13.

Druschel, Harry V.: Analytical characterization of residua and hydrotreated products, Preprints, Amer. Chem. Soc. Div. Petrol. Chem., 1972, 17(4), F92.

Faulk, J. E.: Porphyrins and Metalloporphyrins, Elsevier, Amsterdam, 1964; New edition. K. M. Smith, ed., Elsevier, Amsterdam, 1975.

Gruse, W. A., and D. R. Stevens: Chemical Technology of Petroleum, McGraw-Hill, New York, 1960.

Ho, T. Y., M. A. Rogers, H. V. Druschel, and C. B. Koons: Amer. Assoc. Petrol. Geologist Bull., 1974, 58, 2338.

Koots, J. Anne, and James G. Speight: Relation of petroleum resins to asphaltenes, Fuel, 1975, 54, 179.

Nagy, B., and U. Colombo, eds.: Fundamental Aspects of Petroleum Geochemistry, Elsevier, Amsterdam, 1967.

Rossini, F. D., B. J. Mair, and A. J. Streiff: Hydrocarbons from Petroleum, Reinhold, New York, 1953.

Speight, James G.: The Chemistry and Technology of Petroleum, Marcel Dekker, New York, 1980.

Thompson, C. J., E. M. Shelton, and H. J. Coleman: Sulfur in world crudes, Hydrocarbon Processing, February 1976, 73.

Thompson, C. J., C. C. Ward, and John S. Ball: Characteristics of world's crude oils and results of API Research Project 60, Bartlesville Energy Research Center, Bartlesville, Okla., Report BERC/RI-76/8, July 1976.

van Nes, K., and H. A. van Westen: Aspects of the Constitution of Mineral Oils, Elsevier, Amsterdam, 1951.

2

THE EVALUATION OF HEAVY OILS AND RESIDUA

2-1 REFINERY PROCESSES

In general terms, refinery processes can be conveniently divided into three different types:

1. Separation—i.e., division of the feedstock into various streams (or fractions) depending upon the nature of the crude material
2. Conversion—i.e., the production of salable materials from the feedstock by skeletal alteration, or even by alteration of the chemical type, of the feedstock constituents
3. Finishing—i.e., the purification of the various product streams by a variety of processes which essentially remove impurities from the product

The separation and finishing processes may involve distillation or treatment with a "wash" solution, while the conversion processes are usually regarded as those processes which change the number of carbon atoms per molecule, alter the molecular hydrogen/carbon ratio, or even change the molecular structure of the material (isomerization) without affecting the number of carbon atoms per molecule (Fig. 2-1).

Thus, although it is possible to classify refinery operations in the general terms outlined briefly above, the behavior of various feedstocks in these refinery operations is not so simple. For example, the physical and chemical composition of a feedstock play a large part not only in determining the nature of the products that arise from the refining operations but also in determining the means by which a particular feedstock should be processed. In light of this proviso, it is apparent, therefore that the judicious choice of a crude oil to produce any given product is just as important as the selection of the product for any given purpose. Thus,

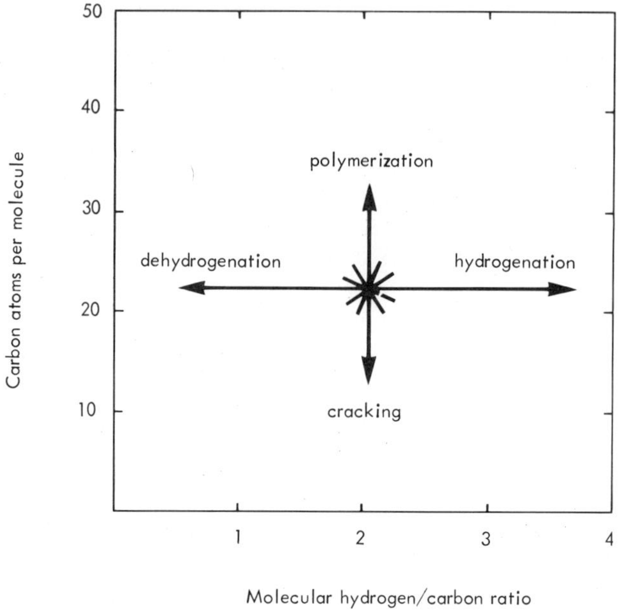

FIG. 2-1 Schematic representation of the chemical changes that occur during the various refinery processes.

initial inspection of the crude oil—conventional examination of the physical properties from which may come deductions about the most logical means of refining—is of the utmost importance. In fact, evaluation of crude oils from physical property data as to which refining sequences should be employed for any particular crude oil is a predominant part of the initial examination of any material that is destined for use as a refinery feedstock.

Thus, the evaluation of crude oil, and for that matter any heavy oil or residuum, usually involves an examination of one or more of the physical properties of the material. As a result, a set of basic characteristics can be obtained which can be correlated with utility and, perhaps what is more important in the present context, with processibility. This also requires comparison of the observed data with those data obtained for other materials of known processibility, as data obtained without any reference point is only of limited value.

In the present context, the heavier feedstocks can be assessed in terms of sulfur content, carbon residue, nitrogen content, and metals content, as well as properties such as the API gravity and viscosity which help the refinery operator to gain an understanding of the nature of the material that is to be processed. Thus, the products from high-sulfur feedstocks often require extensive treatment to remove (or alter) the corrosive sulfur compounds, while nitrogen compounds and the various metals that occur in crude oils will cause serious loss of catalyst life; the carbon residue presents an indication of the amount of coke that may be formed in the process to the detriment of the liquid products.

To satisfy the needs with regard to determining the processibility of a crude oil or a heavier feedstock most refiners have developed their own methods of analysis and evaluation. Although many of these methods are proprietary and, as a result, are not normally available, there are various standards organizations, such as the American Society for Testing and Materials (ASTM) in North America and the Institute of Petroleum (IP) in Britain, which have put considerable time and effort into the correlation and standardization of methods for the inspection and evaluation of refinery feedstocks and refinery products. Accordingly, brief mention is made here of the various tests that might be applied to the inspection and evaluation of a heavy oil or residuum and, for convenience, reference is made to the corresponding ASTM test.

2-2 FEEDSTOCK EVALUATION

A. Sulfur Content

Perhaps the more pertinent property in the present context is the sulfur content, which along with the API gravity represent the two properties which have the greatest influence on the value of a heavy oil and residuum. The sulfur content varies from about 0.1% to about 3% by weight for the more conventional crude oils to as much as 5 to 6% for some of the heavier oils and bitumens (Table 2-1); residua, depending on the sulfur content of the crude oil feedstock, may be of the same order or even have a substantially higher sulfur content. Indeed, the very nature of the distillation process by which residua are produced—i.e., removal of distillate without thermal decomposition—dictates that the majority of the sulfur, which is predominantly in the higher-molecular-weight fractions, be concentrated in the residuum (Figs. 2-2, 2-3, 2-4, and 2-5).

B. Density and Specific Gravity (ASTM D-287, D-1298, D-941, D-1217, and D-1555)

In the early years of the petroleum industry, density was the principal specification for feedstocks and refinery products and was particularly used to give an estimate of the most desirable product, i.e., kerosene, in crude oil. Specific gravity, which approximates closely (but not exactly) to

TABLE 2-1 Elemental Composition of Various Liquid Fossil Fuels

	%					H/C
	C	H	O	N	S	(atomic)
Athabasca bitumen	83.1	10.3	1.3	0.4	4.9	1.487
Cold Lake heavy oil	83.7	10.4	1.1	0.4	4.4	1.491
Qayarah heavy oil	80.7	10.2	—	0.7	8.4	1.519
Crude oil	85.8	13.0	—	0.2	1.0	1.818
Gasoline[a]	86.0	14.0	—	—	—	1.953

[a]Included for comparison.

density, is the more common term at present, but both suffer from the disadvantage that only a very narrow range of values apply to all of the existing crude oils, heavy oils, bitumens, and asphalts. Thus, a conventional crude oil with a high content of paraffins and excellent mobility at ambient temperature and pressure may have a specific gravity (density) of about 0.8, while a heavier asphaltic oil with a high content of asphaltenes and resins and poor mobility at ambient temperature and pressure, thereby requiring vastly different processing sequences, may have a specific gravity (density) of about 0.95 (Table 2-2).

In order to combat this extremely narrow range of values and in an attempt to inject a more meaningful relationship between the physical properties and processibility of the various crude oils, the American Petroleum Institute devised a measurement of gravity devised upon the Baumé scale for industrial liquids. The Baumé scale for liquids lighter than water was used initially:

$$\text{Degrees Baumé} = \frac{140}{\text{sp gr } 60°/60°\text{F}} - 130$$

but a considerable number of hydrometers calibrated according to the Baumé scale were found to be in error by a consistent amount, and this led to modification of the original equation and subsequent adoption of the American Petroleum Institute scale:

$$\text{Degrees API} = \frac{141.5}{\text{sp gr } 60°/60°\text{F}} - 131.5$$

Thus, the API gravity of a lighter crude oil may be of the order of 45° API while a much heavier asphaltic crude oil could have a gravity of the order of 10 to 20° API (Table 2-2) with some residua and natural asphalts (bitumens) having gravities of the order of 5 to 10° API. This is in keeping with the general trend that increased aromaticity leads to a decrease in API gravity (or, more correctly, an increase in specific gravity).

It is also possible to recognize certain preferred trends between the API gravity of petroleums, bitumens, and residua and one or more of the

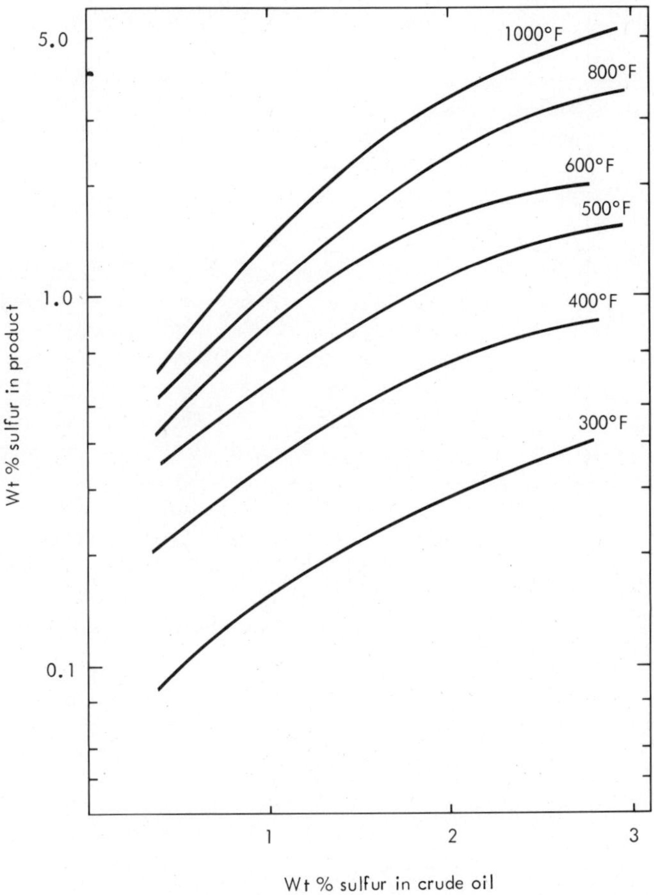

FIG. 2-2 Sulfur content (wt %) of products from West Texas crude oils. (Reproduced with permission from J. H. Gary and G. E. Handwerk, Petroleum Refining: Technology and Economics, 1975, courtesy Marcel Dekker, Inc.)

TABLE 2-2 Physical Characteristics of Some Conventional (Alberta) Crude Oils, Heavy Oils, Bitumens, and Petroleum Residua

Era	Specific gravity	API gravity	Carbon residue, wt/wt %	Sulfur, wt/wt %	Physical composition, wt/wt %		
					Asphaltenes	Resins	Oils
Conventional (Alberta) crude oils[a]							
Upper Cretaceous	0.851	34.8	2.92	0.21	0.9	12.2	86.9
	0.842	36.5	2.98	0.31	0.3	13.0	86.7
	0.850	34.9	2.56	0.25	<0.1	11.6	88.4
	0.830	39.0	1.29	0.16	0.2	9.4	90.4
Lower Cretaceous	0.856	33.8	2.91	0.82	6.9	16.0	77.1
	0.917	22.8	9.32	0.25	11.7	16.5	71.8
	0.908	24.3	8.56	1.71	10.4	16.1	73.5
	0.880	29.3	5.72	1.56	7.2	13.0	79.8
	0.922	22.0	4.45	2.63	15.6	16.7	67.7
	0.810	43.2	0.28	0.08	<0.1	9.4	90.6
	0.837	37.6	1.36	0.15	<0.1	13.7	86.3
	0.851	34.1	2.86	0.19	2.5	6.3	91.2
	0.804	41.4	0.34	0.03	<0.1	6.9	93.1
	0.837	37.5	1.53	0.13	<0.1	9.3	90.7
Mississippian Jurassic	0.894	26.8	4.61	1.05	3.1	12.6	84.3
	0.924	21.6	8.78	2.03	10.9	21.4	67.7
	0.863	32.4	6.25	1.44	3.8	8.4	87.8
	0.849	35.2	0.73	0.46	~0.2	12.5	87.3

	0.822	40.5	1.56	0.22	<0.1	9.1	90.9
	0.905	24.8	6.46	1.80	4.9	10.0	85.1
	0.846	35.7	2.79	0.45	4.9	9.9	85.2
Upper Devonian	0.854	34.2	4.21	1.17	3.4	16.9	79.7
	0.828	39.3	2.59	0.26	0.8	13.7	85.5
	0.829	39.1	1.55	0.35	12.5	12.4	75.1
	0.827	39.8	0.29	0.21	<0.1	3.4	96.6
	0.820	41.1	1.10	0.15	<0.1	5.7	94.3
	0.809	43.4	0.44	0.14	~0.4	10.8	88.8
	0.840	36.9	1.40	0.38	<0.1	7.9	92.1
	0.827	39.7	1.23	0.19	<0.1	6.3	93.7
Lower Devonian	0.890	27.4	6.28	0.88	6.4	14.3	79.3
	0.899	25.8	0.91	1.54	8.1	14.4	77.5
	0.835	37.9	2.77	0.15	2.8	11.1	86.1
	0.825	40.1	2.74	0.24	2.0	9.9	88.1
	0.851	34.7	1.87	0.90	~0.2	12.6	87.2
Heavy oils and natural asphalts (bitumens)							
Cold Lake (Alberta)	0.999	10.1	13.6	4.4	15.7	28.7	55.6
Lloydminister (Alberta)	0.966	15.0	11.8	4.3	12.9	38.4	48.7
Qayarah (Iraq)	0.964	15.3	15.6	8.4	20.4	36.1	43.5
Boscan (Venezuela)	0.998	10.3	10.4[b]	5.6	11.9	24.1	64.0
Athabasca (Canada)	1.030	5.9	18.5	4.9	16.9	34.1	49.0

TABLE 2-2 (Continued)

Era	Specific gravity	API gravity	Carbon residue, wt/wt %	Sulfur, wt/wt %	Physical composition, wt/wt %		
					Asphaltenes	Resins	Oils
Heavy oils and natural asphalts (bitumens) (continued)							
Trinidad Lake	1.070	0.7	10.8[c]	6.2	33.3	29.4	31.9[d]
Bermudez Lake	1.058	2.2	13.4[c]	4.0	35.3	14.4	39.6[d]
Neuchatel (Switzerland)	1.016	7.8	–	–	19.9	32.2	42.2[d]
Selenitza (Albania)	1.080	–	–	6.9	45.2	18.8	24.5
P.R. Springs (Utah)	0.998	10.3	12.5[b]	0.8	16.0	84	
Asphalt Ridge (Utah)	0.986	12.0	9.1[b]	0.4	3.4	96.6	
Tar Sand Triangle	0.992	11.1	21.6[b]	4.4	26.0	74	

Residua							
Wilmington (>485° C; >905° F)	1.010	8.6	18.0[b]	2.1	12.8	87.2	
Mexico[e]	1.015	7.9	28.9[c]	6.4	23.2	76.8	
California[e]	1.065	1.4	24.8[c]	1.2	28.8	71.2	
Gulf Coast[e]	1.022	6.9	25.7[c]	0.8	27.8	72.2	
Venezuela	1.012	8.3	32.0	3.1	11.5	29.3	46.1[d]
Kuwait (>565° C; >1050° F)	1.033	5.5	23.1	5.5	11.1	39.4	49.5

[a] Source: Reproduced from Koots, J. A., and J. G. Speight, Fuel, 1975, 54, 180, by permission of the publisher, IPC Business Press Ltd. ©.

[b] Ramsbottom carbon residue.

[c] Residue on ignition—reported as fixed carbon (mmf).

[d] The remainder appears as asphaltic acids and asphaltic anhydrides.

[e] Cut point not given.

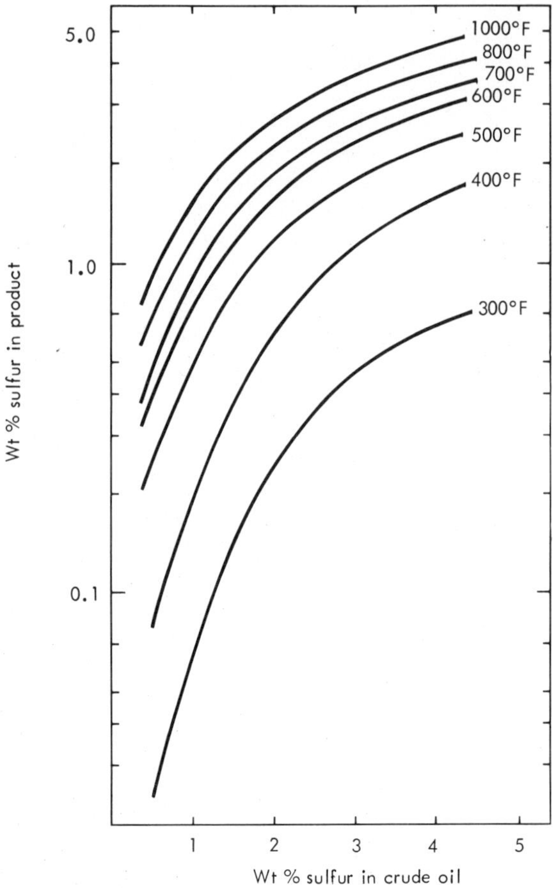

FIG. 2-3 Sulfur content (wt %) of products from California crude oils. (Reproduced with permission from J. H. Gary and G. E. Handwerk, Petroleum Refining: Technology and Economics, 1975, courtesy Marcel Dekker, Inc.)

other physical parameters. For example, correlations exist between the API gravity and sulfur content (Fig. 2-6), Conradson carbon residue (Fig. 2-7), viscosity (Fig. 2-8), asphaltics content (Fig. 2-9), and nitrogen content (Fig. 2-10).

C. Viscosity (ASTM D-445, D-88, D-2161, D-341, and D-2270)

Viscosity is a most important fluid characteristic and indicates the relative mobility of various crude oils and their products, but it is actually a measure of the internal resistance to motion of a fluid by reason of the forces of cohesion between molecules or molecular groupings. There are several means by which viscosity can be measured, with the units of viscosity varying depending upon the method. However, viscosity is often expressed as "stokes" or "poises" (poises = stokes × specific gravity), and the viscosity of crude oils may vary from several stokes (or poises) for the lighter crude oils to several thousand stokes (or poises) for the heavier crude oils even to several million stokes (or poises) for the bitumens and

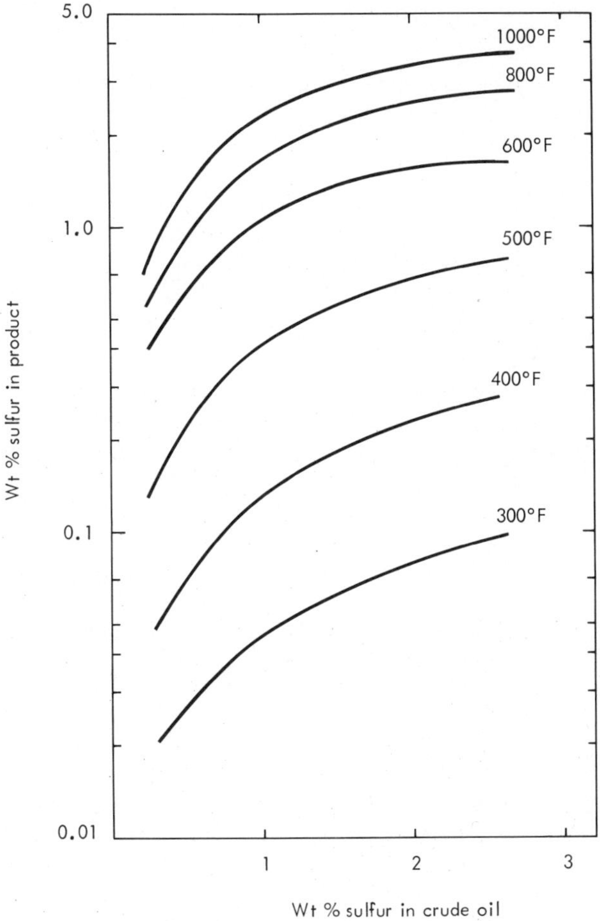

FIG. 2-4 Sulfur content (wt %) of products from various Middle East crude oils. (Reproduced with permission from J. H. Gary and G. E. Handwerk, Petroleum Refining: Technology and Economics, 1975, courtesy Marcel Dekker, Inc.)

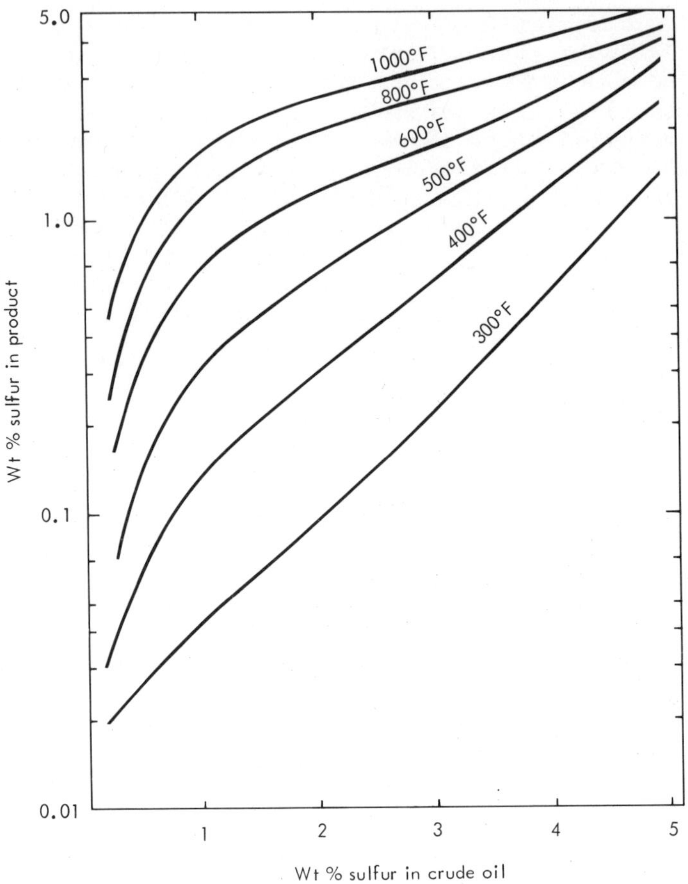

FIG. 2-5 Sulfur content (wt %) of products from various Venezuelan crude oils. (Reproduced with permission from J. H. Gary and G. E. Handwerk, Petroleum Refining: Technology and Economics, 1975, courtesy Marcel Dekker, Inc.)

residua (Figs. 2-11, 2-12, and 2-13). Viscosity data may also be reported as Saybolt universal seconds (SUS) or as Saybolt furol seconds (SFS) which, by means of conversion factors, can be reported as stokes.

The viscosity of heavy oils and residua is an important factor that may have to be taken into account when these materials are recovered from the reservoir or, perhaps what is more pertinent here, when these materials have to be transported from one point in a refinery to another. The high viscosity of certain residua may dictate that the materials are only transported short distances in heated pipes (viscosity tends to decrease markedly with temperature rise) or that the residuum may have to be diluted with an aromatic naphtha which is recovered after the mix reaches its destination.

D. Carbon Residue (ASTM D-189 and D-524)

The carbon residue of a crude oil is a property that can be correlated with the API gravity of the oil and also with the sulfur content, nitrogen content, viscosity, and proportion of asphaltics in the oil (Figs. 2-7 and 2-14; see also Figs. 2-8, 2-9, and 2-10). Hence, the carbon residue presents indications of the "heaviness," or more correctly, the coke-forming propensity of an oil or residuum.

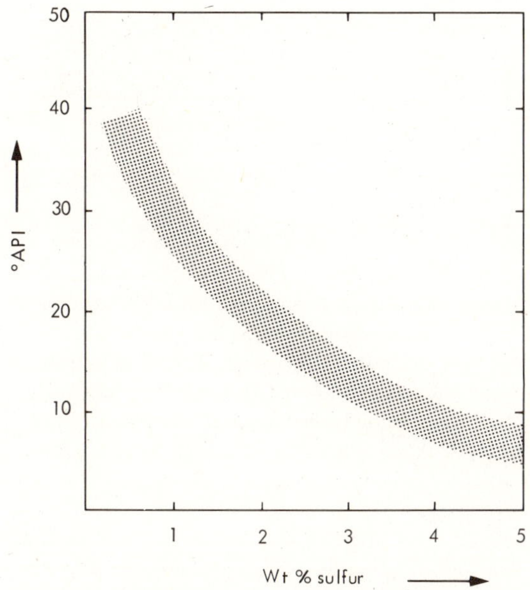

FIG. 2-6 The variation of API gravity with sulfur content of crude oils. (Reproduced with permission from J. G. Speight, The Chemistry and Technology of Petroleum, 1980, courtesy Marcel Dekker, Inc.)

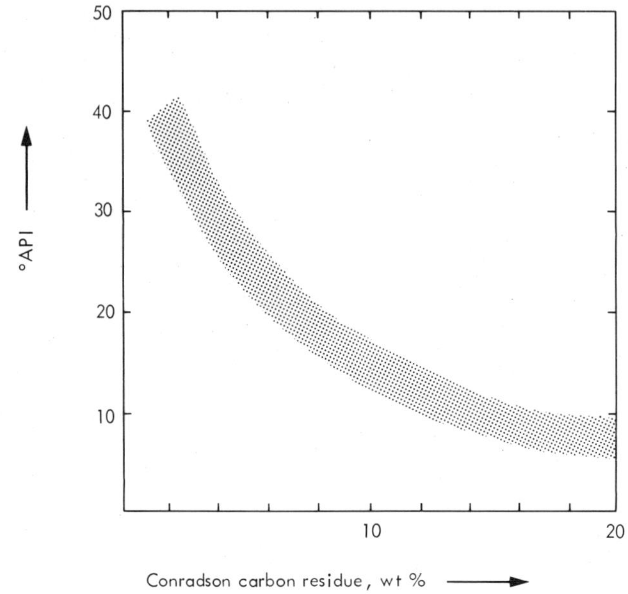

FIG. 2-7 The variation of API gravity with the carbon residue (Conradson) of crude oil. (Reproduced with permission from J. G. Speight, The Chemistry and Technology of Petroleum, 1980, courtesy Marcel Dekker, Inc.)

There are two methods for determining the carbon residue—the Conradson method and the Ramsbottom method—and both are equally applicable to the high-boiling fractions of crude oils which decompose to volatile material and coke when distilled at a pressure of 1 atm. Heavy oils and residua which contain metallic constituents (and distillation of crude oils concentrates these constituents in the residua) will have erroneously high carbon residues. Thus, the metallic constituents must first be removed from the oil or they can be estimated as ash by complete burning of the coke after carbon residue determination.

Although both methods of carbon residue determination involve burning of the sample in a limited amount of oxygen, the techniques employed are quite different; nevertheless, it is possible to interconnect the data (Fig. 2-15). Indeed, it may, for any one particular heavy feedstock and one particular refining operation, be possible to make a rough correlation between carbon residue and product yield. For example, it is generally recognized that during the delayed coking of the bitumen extracted from the Athabasca oil sand, a rise of 1 point (say from 10 to 11) in the Ramsbottom residue of the bitumen feedstock will cause close to a 2 point decrease (say from 77 to 75) in the percentage yield of the coker distillate.

E. Metals Content (ASTM D-482)

The heavier oils and residua contain relatively high proportions of metals (Chapter 1) either in the form of salts or as organometallic constituents (such as the metalloporphyrins) which are extremely difficult to remove from the feedstock. Indeed, the nature of the process by which residua are produced virtually dictates that all of the metals in the original crude oil will be concentrated in the residuum. The exception here is those metallic constituents which may actually volatilize under the distillation conditions and appear in the higher-boiling distillates. The deleterious effect of metallic constituents on the catalyst are known (Chapters 5 and 6), and serious attempts have been made to develop catalysts that can tolerate a high concentration of metals without serious loss in catalyst activity or in catalyst life.

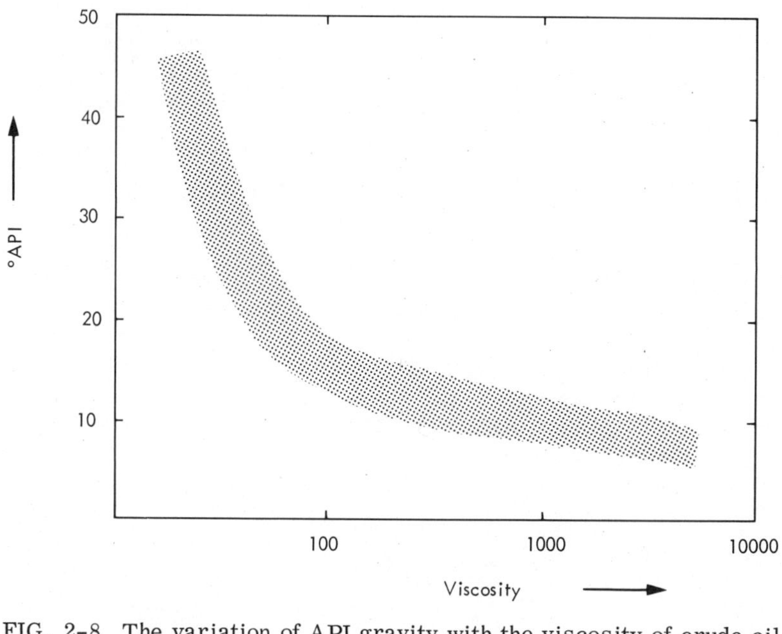

FIG. 2-8 The variation of API gravity with the viscosity of crude oil. (Reproduced with permission from J. G. Speight, The Chemistry and Technology of Petroleum, 1980, courtesy Marcel Dekker, Inc.)

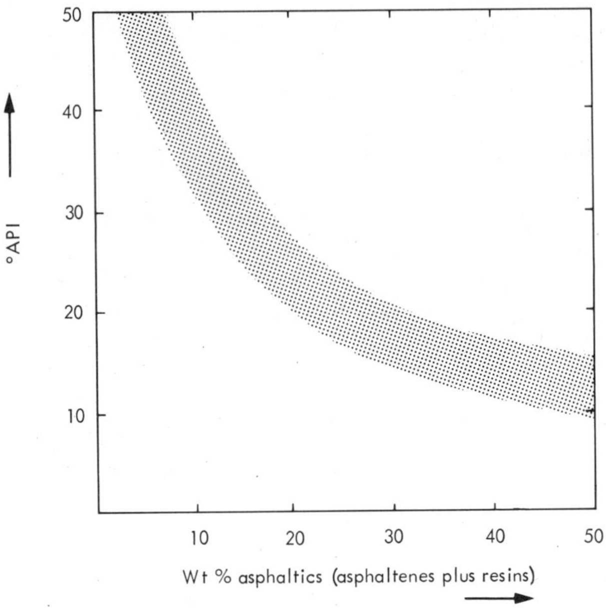

FIG. 2-9 The variation of API gravity with the asphaltics (asphaltenes plus resins) content of crude oil. (Reproduced with permission from J. G. Speight, The Chemistry and Technology of Petroleum, 1980, courtesy Marcel Dekker, Inc.)

The metals content of a feedstock is usually determined as the inorganic ash remaining after ignition and burning of the organic material. Individual metals may then be determined by subjecting the ash to a spectrographic analysis.

F. Distillation Characteristics (ASTM D-86, D-216, D-285, D-477, and D-2892)

The distillation characteristics of crude oils give an indication of the relative proportions of the various products present (Table 2-3). However, for the heavier oils (Table 2-4) the lower-boiling distillation cuts will be present only in very small quantities (if at all) while residua are, by definition, that fraction which remains after the lower-boiling fractions have been removed from a crude oil.

Thus, for a residuum there will be little gain from an examination of the distillation characteristics, while an asphaltic crude oil might be able (according to the distillation characteristics) to produce some of the lower-boiling petroleum products. Nevertheless, the final decision as to whether or not a heavier oil is to be submitted to a distillation sequence in the refinery will be made on the basis of relative amounts of the distillate that will be recovered as well as the overall economics of the various potential processes that can be applied to the oil.

G. Other Properties

Two other properties which may be of value (but only to a lesser extent) to assist in the evaluation of heavy oils and residua are the pour point (ASTM D-97) and the aniline point (ASTM D-611). The pour point of a crude oil product is the lowest temperature at which the oil will pour or

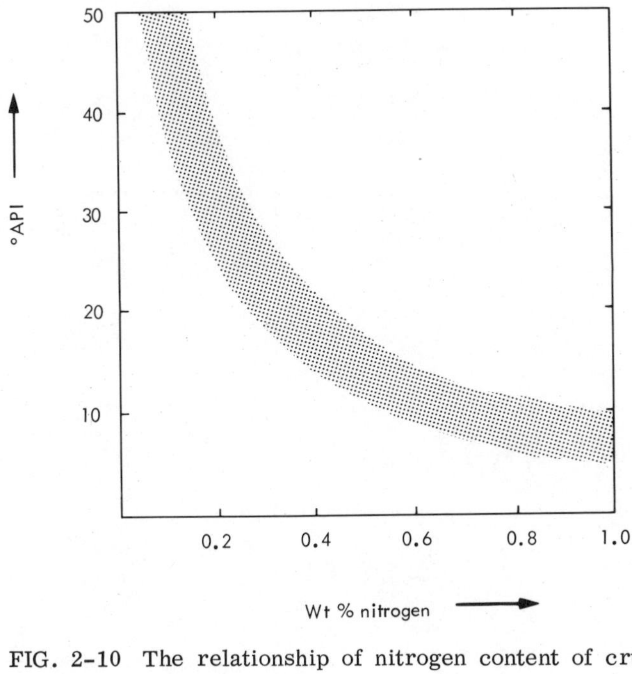

FIG. 2-10 The relationship of nitrogen content of crude oils to API gravity.

TABLE 2-3 Distillation and General Characteristics of a Canadian Conventional Crude Oil[a]

Distillation (U.S. Bureau of Mines Routine Method)

Fraction number	Cut at °C	Cut at °F	%	Sum %	Sp. g. 60° F	°API 60° F	CI	Aniline point °C	SUV 100° F	Cloud test, °F
Stage 1: Distillation at atmospheric pressure, 758 mmHg; first drop, 26° C (79° F)										
1	50	122	2.8	2.8	0.656	84.2	–	–		
2	75	167	3.3	6.1	0.677	77.5	11	56.3		
3	100	212	5.1	11.2	0.713	67.0	18	53.0		
4	125	257	6.8	18.0	0.741	59.5	22	50.6		
5	150	302	6.2	24.2	0.761	54.4	24	50.5		
6	175	347	5.9	30.1	0.779	50.1	26	51.1		
7	200	392	5.0	35.1	0.794	46.7	27	53.7		
8	225	437	5.0	40.1	0.809	43.4	28	58.5		
9	250	482	5.1	45.2	0.821	40.9	29	63.4		
10	275	527	6.1	51.3	0.833	38.4	30	67.7		
Stage 2: Distillation continued at 40 mmHg pressure										
11	200	392	5.2	56.5	0.849	35.2	33	71.3	40	10
12	225	437	4.8	61.3	0.858	33.4	34	76.2	46	30
13	250	482	5.0	66.3	0.870	31.1	36	80.1	57	50
14	275	527	4.7	71.0	0.882	28.9	39	84.0	85	70
15	300	572	4.8	75.8	0.891	27.3	40	88.8	155	90
Residuum			20.2	96.0	0.946	18.1				

Carbon residue of residuum: 6.0%

Approximate Summary

	% by vol.	Sp. g.	°API	SUV 100° F
Light gasoline	11.2	0.688	74.2	
Total gasoline and naphtha	35.1	0.742	59.2	
Kerosine distillate	10.1	0.815	42.1	
Gas oil	15.5	0.845	36.0	
Nonviscous lubricating distillate	9.0	0.862–0.884	32.7–28.6	50–100
Medium lubricating distillate	6.1	0.884–0.895	28.6–26.6	100–200
Viscous lubricating distillate	–	–	–	Above 200
Residuum	20.2	0.946	18.1	
Distillation loss	4.0			

[a] Field: Leduc; specific gravity at 60° F: 0.819; sulfur, wt %: 0.22; Saybolt Universal Viscosity (SUV): at 100° F, sec 39; API gravity at 60° F: 41.3; pour point, °F: 20; color: dark green; carbon residue, wt %: 1.4 (Conradson).

Source: From CANMET Information Circular IC 232, June 1969, 166.

TABLE 2-4 Distillation and General Characteristics of a Canadian Heavy Crude Oil[a]

Distillation (U.S. Bureau of Mines Routine Method)

Fraction number	Cut at °C	Cut at °F	%	Sum %	Sp. g. 60° F	°API 60° F	CI	SUV 100° F	Cloud test, °F
Stage 1: Distillation at atmospheric pressure, 762 mmHg									
1	50	122							
2	75	167							
3	100	212							
4	125	257							
5	150	302	0.9	0.9					
6	175	347	0.8	1.7	0.809	43.4	–		
7	200	392	1.1	2.8	0.823	40.4	41		
8	225	437	1.1	3.9	0.848	35.4	47		
9	250	482	4.1	8.0	0.866	31.9	50		
10	275	527	11.9	19.9	0.867	31.7	46		
Stage 2: Distillation continued at 40 mmHg pressure									
11	200	392	1.6	21.5	0.878	29.7	47	36	Below 0
12	225	437	3.2	24.7	0.929	20.8	67	66	Below 0
13	250	482	6.1	30.8	0.947	17.9	73	118	Below 0
14	275	527	6.4	37.2	0.958	16.2	75	178	Below 0
15	300	572	10.6	47.8	0.972	14.1	78	508	Below 0
Residuum			49.5	97.3					

Carbon residue of residuum: 39.6%
Carbon residue of crude: 19.6%

Approximate Summary				
	% by vol.	Sp. g.	°API	SUV 100° F
Light gasoline	–	–	–	
Total gasoline and naphtha	2.8	0.818	41.5	
Kerosine distillate	–	–	–	
Gas oil	19.0	0.867	31.7	
Nonviscous lubricating distillate	4.3	0.901–0.940	25.6–19.0	50–100
Medium lubricating distillate	8.5	0.940–0.959	19.0–16.1	100–200
Viscous lubricating distillate	13.2	0.959–0.981	16.1–12.7	Above 200
Residuum	49.5	–	–	
Distillation loss	2.7			

[a] Field: McMurray; specific gravity at 60° F: 1.030; sulfur, wt %: 5.84; Saybolt Furol Viscosity: at 210° F, sec 820 and at 250° F, sec 236; API gravity at 60° F: 5.9; pour point, °F: –; color: black; carbon residue, wt %: 19.6 (Conradson)

Source: From CANMET Information Circular IC 232, June 1969, 166. Copyright © Queen's Printer for Canada, 1970, by permission of the publisher.

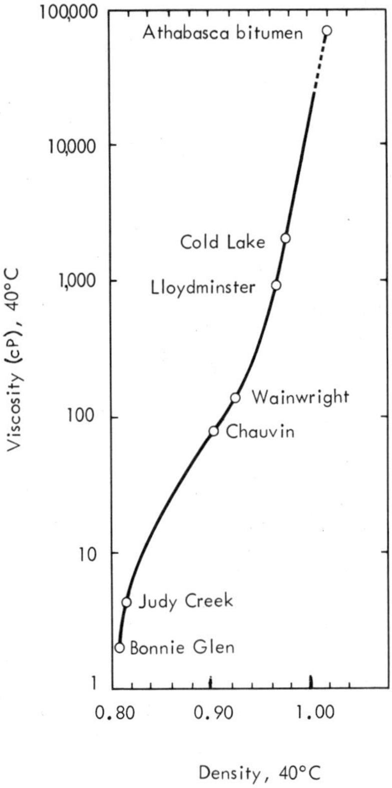

FIG. 2-11 Viscosity-density relationships for Alberta crude oils. (Unpublished data. Reproduced with permission by H. Elmayergi and D. L. Flock.)

flow under prescribed conditions and is a rough indicator of the relative paraffinicity and aromaticity of the material, although for heavy oils and residua pour points are usually high (above 0° C; 32° F) and are more an indication of the temperatures (or conditions) required to move the material "around" the refinery.

The aniline point of a crude oil or crude oil product is the temperature at which equal parts of aniline and the oil are completely miscible. For oils of a given type, the aniline point increases slightly with molecular weight but increases markedly with paraffinicity and may therefore be used to obtain a rough estimation of aromatics content. Aniline point determinations are only infrequently applied to heavy oils and residua since their very character, and the other evaluation methods outlined here, indicates them to be complex with high proportions of ring systems (aromatics and naphthenics).

2-3 USE OF THE DATA

The data derived from any one or more of the evaluation techniques described above can be employed to give the refiner an indication of the means by which the crude feedstock should be processed. However, it must be emphasized that to proceed from the raw evaluation data to full-scale production is not the preferred step; further evaluation of the processibility of the feedstock is usually necessary through the use of a pilot scale operation. To take the evaluation of a feedstock one step further, it may then be possible to develop correlations between the data obtained from the actual plant operations (as well as the pilot plant data) with one or more of the physical properties determined as part of the

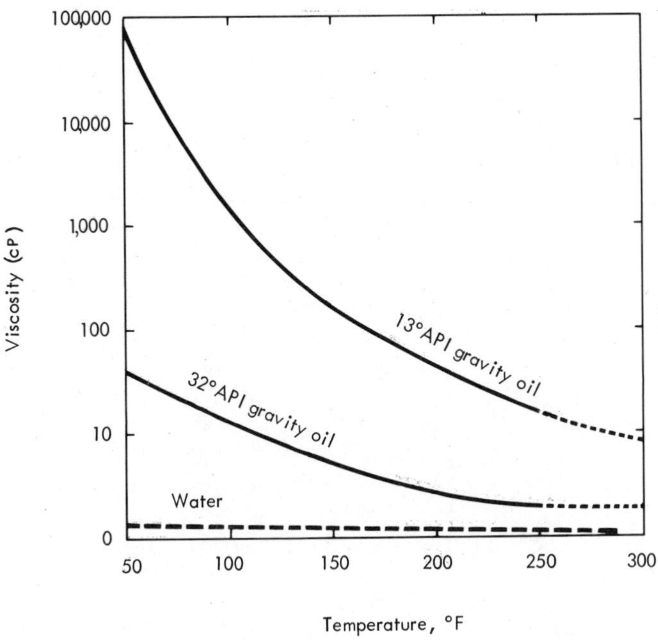

FIG. 2-12 Viscosity-temperature relationships for a conventional crude oil (32° API) and a heavy oil (13° API). (Unpublished data. Reproduced with permission by H. Elmayergi and D. L. Flock.)

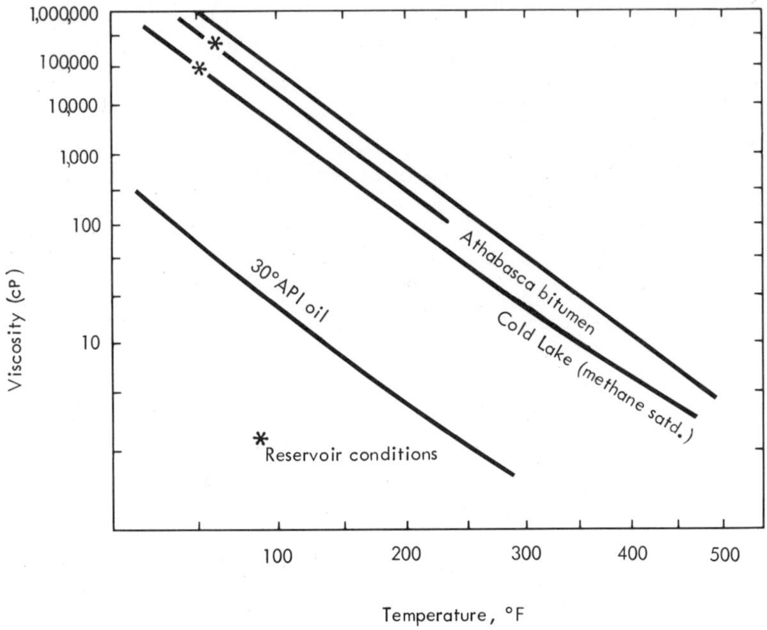

FIG. 2-13 Viscosity-temperature relationships for Alberta bitumen, Alberta heavy oil, and a conventional crude oil.

initial feedstock evaluation. For example, it may be possible to calculate yields for conventional delayed coking operations by using the carbon residue and the API gravity (Table 2-5) of the feedstock.

However, it is essential that when such data are derived, the parameters employed should be carefully specified. For example, the data presented in the tables were derived on the basis of straight-run residua having API gravities less than 18°, the gas oil end point was of the order of 470 to 495°C (875 to 925°F), the gasoline end point was 205°C (400°F), and the pressure in the coke drum was standardized at 35 to 45 psi. Obviously, there are benefits to the derivation of such specific data, but the numerical values, while only representing an approximation, may vary substantially when applied to different feedstocks (Tables 2-6 and 2-7).

TABLE 2-5 Derivation of Product Yields in Delayed Coking from the Carbon Residue (Conradson) and API Gravity of the Feedstock

Coke, wt %	=	$1.6 \times$ wt % Conradson carbon
Gas ($\leqslant C_4$), wt %	=	$7.8 + 0.144$ wt % Conradson carbon
Gasoline, wt %	=	$11.29 + 0.343$ wt % Conradson carbon
Gas oil, wt %	=	100 - wt % coke - wt % gas - wt % gasoline
Gasoline, vol %	=	$\left(\dfrac{186.5}{131.5 + {}^\circ\text{API}}\right)$(gasoline, wt %)
Gas oil, vol %	=	$\left(\dfrac{155.5}{131.5 + {}^\circ\text{API}}\right)$(gas oil, wt %)

Source: Reproduced with permission from J. H. Gary and G. E. Handwerk, Petroleum Refining: Technology and Economics, 1975, courtesy Marcel Dekker, Inc.

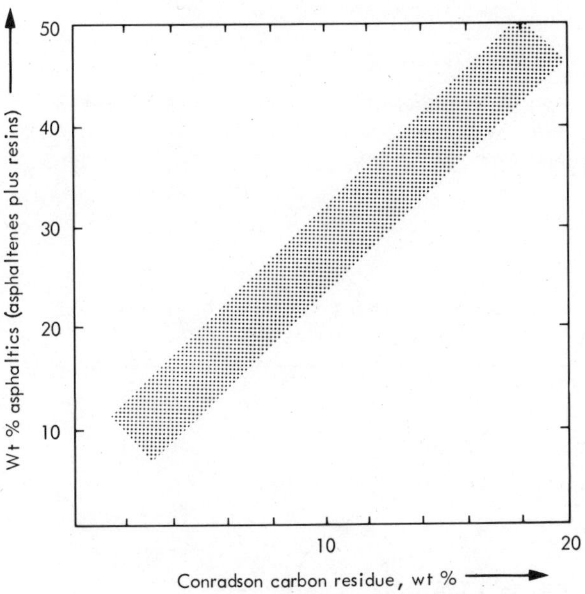

FIG. 2-14 The variation of asphaltics content of crude oils with carbon residue (Conradson). (Reproduced with permission from J. G. Speight, The Chemistry and Technology of Petroleum, 1980, courtesy Marcel Dekker, Inc.)

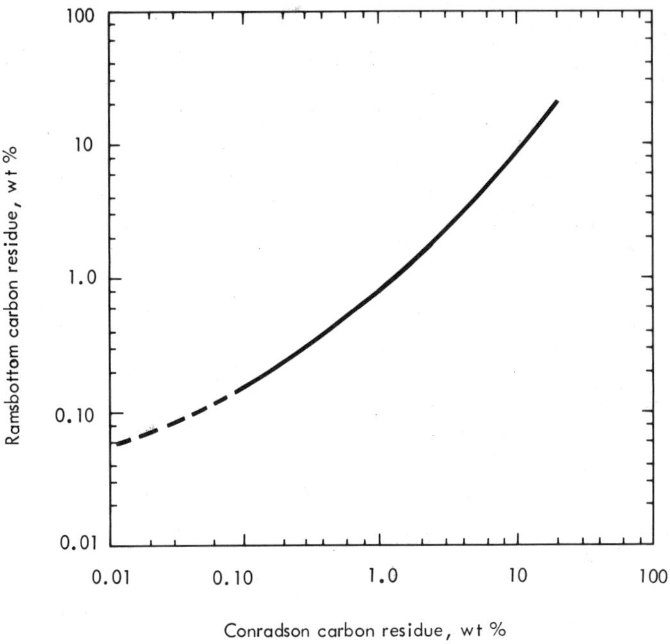

FIG. 2-15 Interconversion of Ramsbottom and Conradson carbon residue data. (Reprinted with permission from the 1979 Annual Book of ASTM Standards, Part 23, p. 128. Copyright American Society for Testing and Materials, 1916 Race Street, Philadelphia, Pa. 19103.)

TABLE 2-6 Product Yields in the Delayed Coking of Wilmington Straight-Run Residuum

$$\text{Coke, wt \%} = 39.68 - 1.60 \times {}^\circ\text{API}$$

$$\text{Gas } (\leqslant C_4)\text{, wt \%} = 11.27 - 0.14 \times {}^\circ\text{API}$$

$$\text{Gasoline, wt \%} = 20.5 - 0.36 \times {}^\circ\text{API}$$

$$\text{Gas oil, wt \%} = 28.55 + 2.10 \times {}^\circ\text{API}$$

$$\text{Gasoline, vol \%} = \left(\frac{186.5}{131.5 + {}^\circ\text{API}}\right)(\text{gasoline, wt \%})$$

$$\text{Gas oil, vol \%} = \left(\frac{155.5}{131.5 + {}^\circ\text{API}}\right)(\text{gas oil, wt \%})$$

Source: Reproduced with permission from J. H. Gary and G. E. Handwerk, Petroleum Refining: Technology and Economics, 1975, courtesy Marcel Dekker, Inc.

TABLE 2-7 Product Yields in the Delayed Coking of East Texas Straight-Run Residuum

$$\text{Coke, wt \%} = 45.76 - 1.78 \times {}^{\circ}\text{API}$$

$$\text{Gas } (\leqslant C_4)\text{, wt \%} = 11.92 - 0.16 \times {}^{\circ}\text{API}$$

$$\text{Gasoline, wt \%} = 20.5 - 0.36 \times {}^{\circ}\text{API}$$

$$\text{Gasoline, vol \%} = \left(\frac{186.5}{131.5 + {}^{\circ}\text{API}}\right)(\text{gasoline, wt \%})$$

$$\text{Gas oil, vol \%} = \left(\frac{155.5}{131.5 + {}^{\circ}\text{API}}\right)(\text{gas oil, wt \%})$$

Source: Reproduced with permission from J. H. Gary and G. E. Handwerk, Petroleum Refining: Technology and Economics, 1975, courtesy Marcel Dekker, Inc.

BIBLIOGRAPHY

Abraham, Herbert: Asphalts and Allied Substances, 5th ed., Vol. 1, Van Nostrand, New York, 1945.

American Society for Testing and Materials: 1979 Annual Book of ASTM Standards, Philadelphia, 1979.

Bland, William F., and Robert L. Davidson, eds.: Petroleum Processing Handbook, McGraw-Hill Book Company, New York, 1967.

Gary, James H., and Glenn E. Handwerk: Petroleum Refining: Technology and Economics, Marcel Dekker, New York, 1975.

Nelson, Wilbur L.: Petroleum Refinery Engineering, McGraw-Hill, New York, 1958.

Speight, James, G.: The Chemistry and Technology of Petroleum, Marcel Dekker, New York, 1980.

3

THE CHEMISTRY AND PHYSICS OF DESULFURIZATION

3-1 CHEMICAL CONCEPTS

Sulfur removal, as currently practiced in the petroleum industry, can be achieved by (1) thermal methods such as the various cracking techniques which essentially concentrate the majority of the sulfur into the nonvolatile products, i.e., the pitches and cokes, while the remainder of the sulfur may occur in the gases and as low-boiling organic sulfur compounds; (2) chemical methods such as alkali treating (sweetening) as may be located in the product "finishing" section of a refinery; and (3) hydrodesulfurization. It is this latter process that will receive the major emphasis throughout this text in the manner in which it relates to the processing of heavy oils and residua.

Hydrodesulfurization is a catalytic process whereby a crude oil fraction (or product), as the feestock, is passed, with hydrogen, over or through a catalyst bed at elevated temperatures and pressures. The basic chemical concept of the process is to convert the organic sulfur in the feedstock to hydrogen sulfide. The hydrodesulfurization process is one of the processes that may be generally referred to as hydrogenation processes.

Hydrogenation processes for the conversion of crude oil fractions and products may be classified as nondestructive and destructive. Although the definition of the two processes is purely arbitrary, it is generally assumed that destructive hydrogenation (which is characterized by the cleavage of carbon-to-carbon linkages and is accompanied by hydrogen saturation of the fragments to produce lower-boiling products) requires temperatures in excess of 350°C (660°F). However, nondestructive hydrogenation is more generally used for the purpose of improving product quality without any appreciable alteration of the boiling range. Mild processing conditions (temperatures below 350°C or 660°F) are employed so that only the more unstable materials are attached. Thus, sulfur, nitrogen, and oxygen compounds eliminate hydrogen sulfide, ammonia, and water, respectively, and unsaturated materials are hydrogenated to produce the more stable hydrocarbons.

While the definitions of the various hydroprocesses are (as has been noted above) quite arbitrary, it may be difficult, if not impossible, to limit the process to any one particular reaction in a commercial operation, but the prevailing conditions may, to a certain extent, minimize, say, cracking reactions during a hydrotreating operation. However, with respect to the heavier feedstocks, the ultimate aim of the operation is to produce as much low-sulfur liquid products as possible from the feedstock and, thus, any hydrodesulfurization process that has been designed for application to the heavier oils and residua may require that hydrocracking and hydrodesulfurization occur simultaneously.

Under the usual commercial hydrodesulfurization conditions (elevated temperatures and pressures, high hydrogen-to-feedstock ratios, and the presence of a catalyst), the various reactions that result in the removal of sulfur from the organic feedstock (Table 3-1) occur. Thus, thiols and open-chain and cyclic sulfides are converted to saturated and/or aromatic compounds—depending, of course, on the nature of the particular sulfur compound involved. Benzothiophenes are converted to alkylated aromatics, while dibenzothiophenes are usually converted to biphenyls. In fact, the major reactions which occur as part of the hydrodesulfurization process involve simple carbon-sulfur bond rupture and saturation of the reactive fragments (as well as saturation of olefinic material). Aromatic rings are not usually saturated (even though hydrogenation of the aromatic rings may be thermodynamically favored) but the saturation of some of the aromatic rings may appear to have occurred because of the tendency of partial ring saturation to occur <u>prior</u> to carbon-sulfur bond rupture as is believed to happen in the case of dibenzothiophene derivatives. It is generally recognized that the ease of desulfurization is dependent upon the type of compound, and the lower-boiling fractions are desulfurized more easily than the higher-boiling fractions. The difficulty of sulfur removal increases in the order:

paraffins $<$ naphthenes $<$ aromatics

The wide ranges of temperature and pressure employed for the hydrodesulfurization process virtually dictate that many other reactions will proceed concurrently with the desulfurization reaction. Thus, the isomerization of paraffins and naphthenes may occur and hydrocracking will increase as the temperature and pressure increase. Furthermore, at the higher temperatures (but low pressures) naphthenes may dehydrogenate to aromatics and paraffins dehydrocyclize to naphthenes, while at lower temperatures (high pressures) some of the aromatics may be hydrogenated. That these reactions do not all occur equally is, to some extent, also due

TABLE 3-1 Typical Hydrodesulfurization Reactions

1. Mercaptans:

$$R{-}SH + H_2 \longrightarrow RH + H_2S$$

2. Disulfides:

$$R{-}SS{-}R' + 3H_2 \longrightarrow RH + R'H + 2H_2S$$

3. Sulfides (aromatic, naphthenic, and alkyl):

$$R{-}S{-}R' + 2H_2 \longrightarrow RH + R'H + H_2S$$

4. Thiophenes:

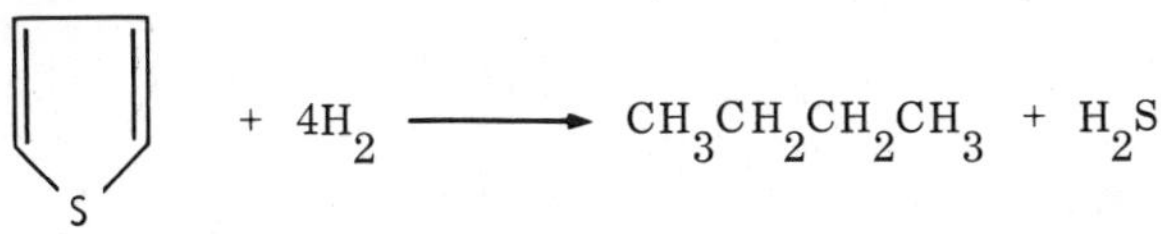

5. Benzothiophenes:

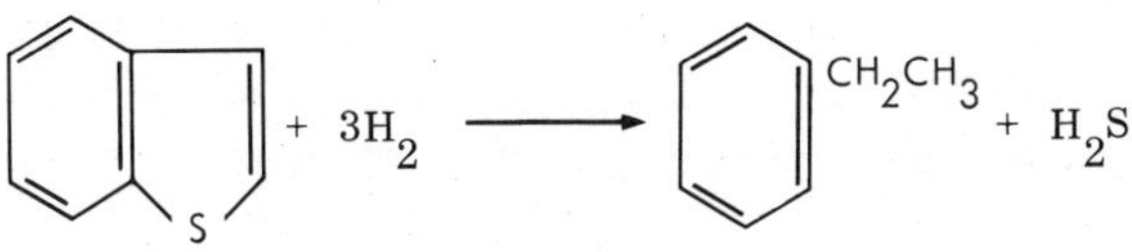

6. Dibenzothiophenes:

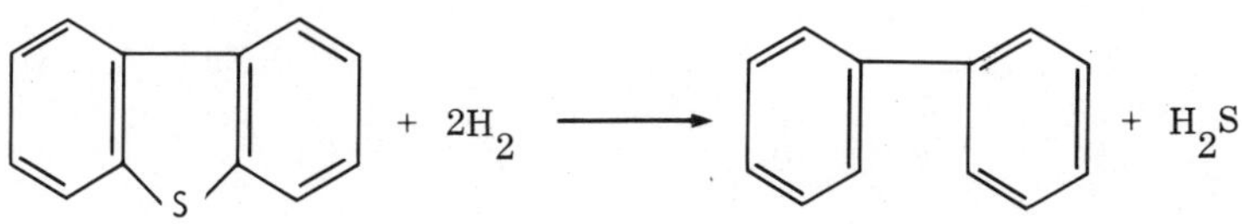

to the nature of the catalyst— judicious choice of a catalyst will lead to the elimination of sulfur (and the other heteroatoms nitrogen and oxygen) and, although some hydrogenation and hydrocracking may occur, the extent of such reactions (Table 3-2) may be relatively minor. In the present context, the hydrodesulfurization of the heavier feedstocks (heavy oils and residua)

TABLE 3-2 Typical Reactions that Might Occur During a Hydroprocessing Operation

1. Denitrogenation:

   (a) Pyrrole

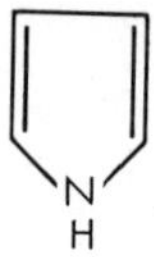 $+ 4H_2 \longrightarrow CH_3CH_2CH_2CH_3 + NH_3$

   (b) Pyridine

 $+ 5H_2 \longrightarrow CH_3CH_2CH_2CH_2CH_3 + NH_3$

2. Deoxygenation:

   (a) Phenol

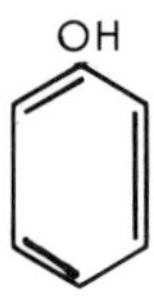 $+ H_2 \longrightarrow$  $+ H_2O$

   (b) Furan

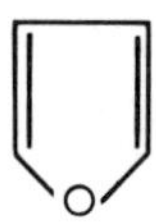 $+ 4H_2 \longrightarrow CH_3CH_2CH_2CH_3 + H_2O$

3. Hydrogenation:

$$CH_3CH_2CH_2CH{:}CH_2 + H_2 \longrightarrow CH_3CH_2CH_2CH_2CH_3$$

4. Hydrocracking:

$$C_{10}H_{22} + H_2 \longrightarrow C_4H_{10} + C_6H_{14}$$

may require that part (or almost all) of the feedstock be converted to lower-boiling products. If this be the case (as is now usual), hydrocracking will, of course, compete on an almost equal footing with the hydrodesulfurization reaction for the production of the low-sulfur, low-boiling products.

Thus, the hydrodesulfurization process is a very complex sequence of reactions due, no doubt, to the complexity of the feedstock. Furthermore, the fact that feedstocks usually contain nitrogen and oxygen compounds (in addition to metal compounds) increases the complexity of the reactions which occur as part of the hydrodesulfurization process. The nitrogen compounds which may be present are typified by pyridines, quinolines, carbazoles, indoles, and pyrroles (Chapter 1). Oxygen may be present as phenols and carboxylic acids (Chapter 1); the most common metals to occur in petroleums are nickel and vanadium (Chapter 1). In the majority of conventional petroleums, these other atoms (nitrogen, oxygen, and metals) may be of little consequence, but heavy oils contain substantial amounts of these atoms, and the nature of the process by which residua arise dictates that virtually all of the metals and substantial amounts of the nitrogen and oxygen originally present in the petroleum will be concentrated in the residua.

The simultaneous removal of nitrogen during the processing is a very important aspect of hydrodesulfurization. Compounds containing nitrogen have pronounced deleterious effects on the storage stability of petroleum products, and nitrogen compounds in charge stocks to catalytic processes can severely limit (and even poison) the activity of the catalyst. Oxygen compounds are corrosive (especially the naphthenic acids) and can promote gum formation as part of the deterioration of the hydrocarbons in the petroleum product. Metals in feedstocks that are destined for catalytic processes can poison the selectivity of the catalyst and, like the nitrogen and oxygen compounds, should be removed. Fortunately, the continued developments of the hydrodesulfurization process over the last two decades has resulted in the production of catalysts which can tolerate substantial amounts of nitrogen compounds, oxygen compounds, and metals without serious losses in catalyst activity or in catalyst life (Chapter 5). Thus, it is possible to use the hydrodesulfurization process not only as a means of producing low-sulfur liquid products but also as a means of producing low-sulfur, low-nitrogen, low-oxygen, and low-metals streams that can be employed as feedstocks for processes where catalyst sensitivity is one of the process features.

3-2 THERMODYNAMICS OF HYDRODESULFURIZATION

The thermodynamics of the hydrodesulfurization reaction has been evaluated from the equilibrium constants of typical desulfurization or partial desulfurization reactions such as:

1. Hydrogenation of model compounds to yield saturated hydrocarbons and hydrogen sulfide
2. Decomposition of model compounds to yield unsaturated hydrocarbons and hydrogen sulfide
3. Decomposition of alkyl sulfides to yield thiols and olefins
4. Condensation of thiols to yield alkyl sulfides and hydrogen sulfide
5. Hydrogenation of disulfides to yield thiols

The logarithms of the equilibrium constants for the reduction of sulfur compounds to saturated hydrocarbons over a wide temperature range (Table 3-3) are almost all positive, indicating that the reaction can virtually proceed to completion if hydrogen is present in the stoichiometric quantity. The equilibrium constant does, however, decrease with increasing temperature for each particular reaction but still does retain a substantially positive value at 425° C (795° F) which is approaching the maximum temperature at which many of the hydrodesulfurization (especially nondestructive) reactions would be attempted. The data also indicate that the decomposition of sulfur compounds to yield unsaturated hydrocarbons and hydrogen sulfide is not thermodynamically favored at temperatures below 325° C (615° F) and such a reaction has no guarantee of completion until temperatures of about 625° C (1155° F) are reached. However, substantial decomposition of thiols can occur at temperatures below 300° C (570° F); in fact (with only few exceptions), the decomposition of all saturated sulfur compounds is thermodynamically favored at temperatures below 425° C (795° F).

The data generally indicate the types of reactions which can occur during the hydrodesulfurization reaction and include those reactions which will occur at the upper end of the temperature range of the hydrodesulfurization process whether it be a "true" hydrodesulfurization reaction or a cracking reaction. Even though some of the reactions given here may only be incidental, they must nevertheless be taken into account because of the complex nature of the feedstock and the several process variations (Chapter 5) which (in addition to the fact that the overall hydrodesulfurization process is exothermic—Table 3-4) all contribute to the complexity of the product mix.

3-3 KINETICS OF HYDRODESULFURIZATION

Kinetic studies using individual sulfur compounds have usually indicated that simple first-order kinetics with respect to sulfur is the predominant mechanism by which sulfur is removed from the organic material as hydrogen sulfide. However, there is still much to be learned about the relative rates

TABLE 3-3 Logarithms of Equilibrium Constants for the Reactions of Sulfur Compounds (see also Figs. 3-1 and 3-2)

Reaction	$\log_{10} K_{eq}$ at 298.16	500	700	900° K
Sulfur compound + H_2 → saturated hydrocarbon + H_2S:				
Methanethiol → methane	+14.28	+ 8.37	+ 6.10	+ 4.69
Ethanethiol → ethane	+11.70	+ 7.06	+ 5.01	+ 3.84
2-Propanethiol → propane	+ 9.59	+ 6.05	+ 4.45	+ 3.52
2-Methyl-2-propanethiol → isobutane	+ 9.43	+ 6.18	+ 4.68	+ 3.81
1-Pentanethiol → n-pentane	+10.39	—	—	—
2-Thiapropane → methane	+25.45	+15.68	+11.41	+ 8.96
2-Thiabutane → methane + ethane	+22.58	+13.85	+ 9.97	+ 7.74
3-Thiapentane → ethane	+20.08	+12.52	+ 9.11	+ 7.13
3-Methyl-2-thiabutane → methane + propane	+21.30	+13.37	+ 9.80	+ 7.72
Thiacyclopropane → ethane	+28.32	+15.02	+ 9.26	+ 5.99
Thiacyclobutane → propane	+28.88	+15.61	+ 9.90	+ 6.67
Thiacyclopentane → n-butane	+16.83	+ 8.79	+ 5.26	+ 3.24
Thiacyclohexane → n-pentane	+16.67	+ 9.22	+ 5.92	+ 3.97
2, 3-Dithiabutane → methane	+42.21	+26.08	+19.03	+14.97
3, 4-Dithiahexane → ethane	+36.75	+22.94	+16.79	+13.23
Thiophene → n-butane	+30.89	+12.07	+ 3.85	- 0.85
3-Methylthiophene → 2-methylbutane	+29.85	+11.27	+ 3.17	- 1.43
2-Methylthiophene → n-pentane	+28.91	+10.84	+ 2.94	- 1.55
Sulfur compound → unsaturated hydrocarbon + H_2S:				
Ethanethiol → ethylene	- 5.98	- 0.83	+ 1.34	+ 2.53
1-Propanethiol → propylene	—	+ 0.18	—	—
2-Propanethiol → propylene	- 5.51	- 0.16	+ 2.06	+ 3.26
1-Pentanethiol → 1-pentene	- 4.97	—	—	—
2-Methyl-2-propanethiol → isobutylene	- 4.40	+ 0.76	+ 2.89	+ 4.04
3-Thiapentane → ethylene	-15.30	- 3.28	+ 1.75	+ 4.53
Thiacyclopropane → acetylene	-14.09	- 5.10	- 1.22	+ 0.92
Thiacyclobutane → allene	-10.68	- 2.79	+ 0.61	+ 2.50
Thiacyclobutane → propyne	-10.63	- 4.73	- 2.23	- 0.88
Thiacyclopentane → 1, 3-butadiene	-12.53	- 3.67	+ 0.14	+ 2.26
Thiacyclopentane → 1-butyne	-21.54	- 8.79	- 3.35	- 0.35

TABLE 3-3 (Continued)

Reaction	$\log_{10} K_{eq}$ at 298.16	500	700	900° K
Sulfur compound → unsaturated hydrocarbon + H_2S (continued)				
Thiacyclohexane → 1,4-pentadiene	-14.60	- 4.04	+ 0.46	+ 2.91
Thiacyclohexane → 1-pentyne	-21.56	- 8.23	- 2.57	+ 0.49
3,4-Dithiahexane → ethylene + acetylene	-23.33	- 5.08	+ 2.64	+ 6.85
Open-chain sulfide → thiol + olefin:				
2-Thiabutane → methanethiol + ethylene	- 9.38	- 2.63	+ 0.20	+ 1.75
3-Thiapentane → ethanethiol + ethylene	- 9.32	- 2.45	+ 0.41	+ 2.00
3-Methyl-2-thiabutane → methanethiol + propylene	- 8.08	- 1.42	+ 1.31	+ 2.79
Thiol → open-chain sulfide + H_2S:				
Methanethiol → 2-thiapropane	+ 3.10	+ 1.47	+ 0.79	+ 0.41
Methanethiol + ethanethiol 2-thiabutane	+ 3.40	+ 1.80	+ 1.14	+ 0.78
Ethanethiol → 3-thiapentane	+ 3.34	+ 1.70	+ 0.93	+ 0.53
Methanethiol + 2-propanethiol 3-methyl-2-thiabutane	+ 2.57	+ 1.26	+ 0.75	+ 0.47
Open-chain disulfide + H_2 → thiol:				
2,3-Dithiabutane → methanethiol	+13.67	+ 8.93	+ 6.83	+ 5.62
3,4-Dithiahexane → ethanethiol	+13.35	+ 8.80	+ 6.77	+ 5.56

Source: Reproduced with permission from J. B. McKinley, Catalysis, P. H. Emmett, ed., 1957, Vol. 5, 405.

of reaction exhibited by the various compounds present in petroleum. Nevertheless, there are several generalizations which come from the available thermodynamic data and investigations of pure compounds as well as work carried out on petroleum fractions. Thus, at room temperature, hydrogenation of sulfur compounds to hydrogen sulfide is thermodynamically favorable and the reaction will essentially proceed to completion in the presence of a stoichiometric amount of hydrogen. Sulfides, simple thiophenes, and benzothiophenes are generally easier to desulfurize than the dibenzothiophenes and the higher-molecular-weight condensed thiophenes.

However, the development of generalized kinetic data for the hydrodesulfurization of different petroleum feedstocks is complicated by the presence of a large number of sulfur compounds each of which may react at a different rate because of structural differences as well as differences in molecular weight. In addition, the sulfur compounds in a feedstock may cause changes in the catalyst upon contact and, therefore, every effort should be made to ensure that the kinetic data from such investigations are derived under a set of standardized conditions. In this sense, several attempts have been made to accomplish standardization of the reaction conditions by presulfiding the catalyst, passage of the feedstock over the catalyst until the catalyst is "stabilized," obtaining the data at various conditions, and then rechecking the initial data by repetition.

Thus, it has become possible to define certain general trends that occur in the hydrodesulfurization of petroleum feedstocks. One of the more noticeable facets of the hydrodesulfurization process is that the rate of reaction declines markedly with the molecular weight of the feedstock (Fig. 3-3)—for example, examination of the thiophenic portion of a

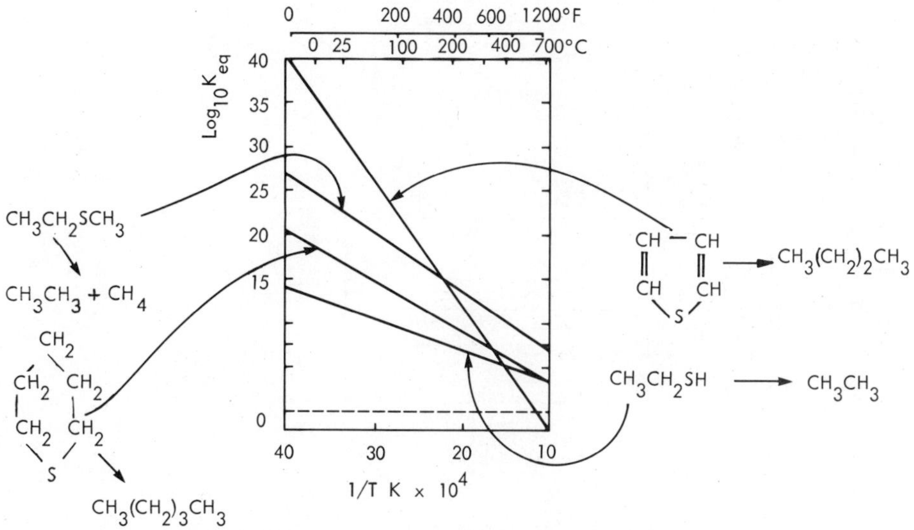

FIG. 3-1 Thermodynamic data for the hydrodesulfurization of model sulfur compounds (data from Table 3-3).

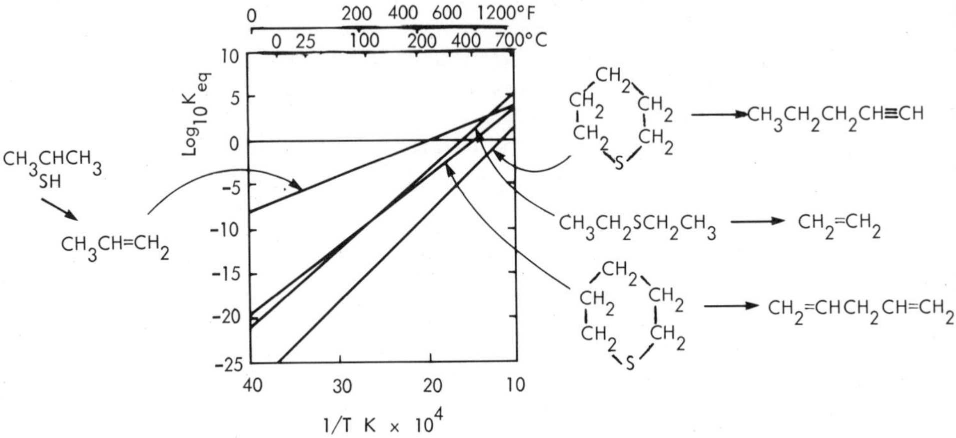

FIG. 3-2 Thermodynamic data for the desulfurization of model sulfur compounds (data from Table 3-3).

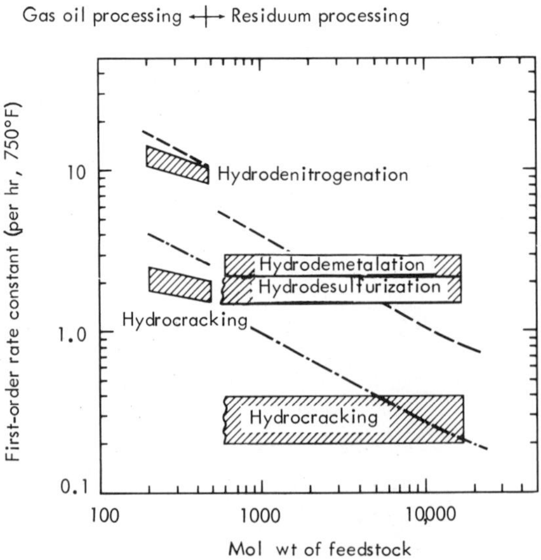

FIG. 3-3 Effect of molecular weight of the feedstock on hydroconversion processes. [From J. W. Scott and A. G. Bridge, Origin and refining of petroleum, in Advances in Chemistry Series, No. 103 (H. G. McGrath and M. E. Charles, eds.), American Chemical Society, Washington, D.C., 1971, p. 123.]

TABLE 3-4 Heats of Reaction for Various Model Hydrodesulfurization Reactions

Reaction	$\Delta H^\circ_{700\ K}$ (kcal mol^{-1})
$C_2H_5SH + H_2 \rightarrow C_2H_6 + H_2S$ (Ethanethiol)	-16.77
$C_2H_5SC_2H_5 + 2H_2 \rightarrow 2C_2H_4 + H_2S$ (Diethyl sulfide)	-27.99
$C_4H_8S + 2H_2 \rightarrow n\text{-}C_4H_{10} + H_2S$ (Thiacyclopentane)	-28.73
$C_4H_4S + 4H_2 \rightarrow n\text{-}C_4H_{10} + H_2S$ (Thiophene)	-66.98
$CH_3C_4H_3S + 4H_2 \rightarrow C_5H_{12} + H_2S$ (3-Methylthiophene) (2-Methyl butane)	-65.79
$C_6H_{12} + H_2 \rightarrow n\text{-}C_4H_{14}$ (Hexene-1)	-31.10

(narrow-boiling) feedstock and the resulting hydrodesulfurized product provides excellent evidence that benzothiophenes are removed in preference to the dibenzothiophenes and other condensed thiophenes—the ratio of higher condensed thiophenes to benzothiophenes was about 2:1, whereas in the hydrodesulfurized product this ratio was of the order of 20:1; the sulfur compounds in residua (and heavy oils) are presumed to react (preferentially) in a similar manner.

It is also generally accepted that the simpler sulfur compounds (e.g., thiols and thioethers) are (unless steric influences offer resistance to the hydrodesulfurization) easier to remove from petroleum feedstocks than the more complex sulfur compounds (e.g., thiophenes). It should be noted here that, because of the nature of the reaction, steric influences would be anticipated to play a lesser role in the hydrocracking process.

3-4 KINETICS OF RESIDUA HYDRODESULFURIZATION

Residua hydrodesulfurization is considerably more complex than the hydrodesulfurization of model organic sulfur compounds or, for that matter, narrow-boiling petroleum fractions. In published studies of the kinetics of residua hydrodesulfurization, one of three approaches has generally been taken:

1. The reactions can be described in terms of simple first-order expressions.
2. The reactions can be described by use of two simultaneous first-order expressions—one expression for easy-to-remove sulfur and a separate expression for difficult-to-remove sulfur.
3. The reactions can be described using a pseudo-second-order treatment.

Each of the three approaches has been used to describe hydrodesulfurization of residua under a variety of conditions with varying degrees of success, but it does appear that pseudo-second-order kinetics are favored. In this particular treatment, the rate of hydrodesulfurization is expressed by a simple second-order equation:

$$\frac{C}{1 - C} = k \frac{1}{\text{LHSV}}$$

where

$$C = \frac{\text{wt \% sulfur in product}}{\text{wt \% sulfur in charge}}$$

k = reaction rate constant

LHSV = liquid hourly space velocity (volume of liquid feed per hour per volume of catalyst)

Indeed, application of this model to a Kuwait vacuum residuum (18% by weight of the crude oil) using a conventional desulfurization catalyst and a catalyst of greatly improved desulfurization activity (Fig. 3-4) gave a linear relationship. However, it is somewhat difficult to visualize the desulfurization reaction as a reaction which requires the interaction of two sulfur-containing molecules (as dictated by the second-order kinetics). To accommodate this anomaly, it has been suggested that, as there are many different types of sulfur compounds in residua and each may react at a different rate, the differences in reaction rates offered a reasonable explanation for the apparent second-order behavior. For example, an investigation of the hydrodesulfurization of an Arabian light-atmospheric residuum

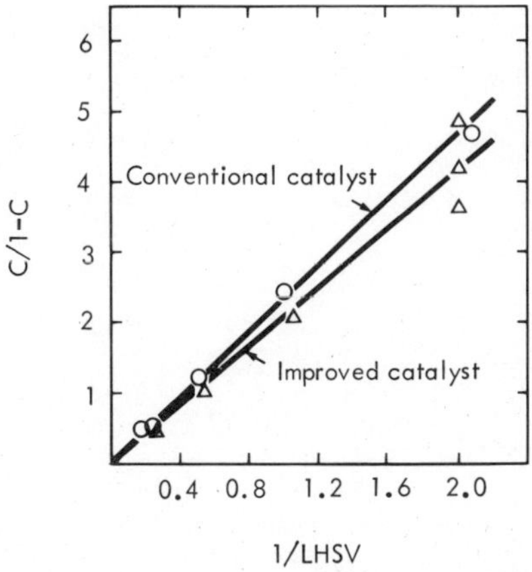

FIG. 3-4 Hydrodesulfurization of a vacuum residuum (Kuwait) as a second-order reaction. (From H. Beuttner and B. K. Schmid, Proc. 6th World Petroleum Congress, 1963, Vol. 3, p. 297.)

showed that the overall reaction could not be adequately represented by a first-order relationship (Fig. 3-5). However, the reaction could be represented as the sum of two competing first-order reactions and the rates of desulfurization of the two fractions (the oil fraction and the asphaltene fraction) could be well represented as an overall second-order reaction.

Thus, if each type of sulfur compound is removed by a reaction that was first-order with respect to sulfur concentration, the first-order reaction rate would gradually, and continually, decrease as the more reactive sulfur compounds in the mix became depleted, leaving the more stable and hence more-difficult-to-remove sulfur compounds in the residue. This sequence of events will, presumably, lead to an apparent second-order rate equation which is, in fact, a compilation of many consecutive first-order reactions of continually decreasing rate constant. Indeed, the desulfurization of model sulfur-containing compounds exhibits first-order kinetics, and the concept that the residuum consists of a series of first-order reactions of decreasing rate constant leading to an overall second-order effect has been found to be acceptable.

Application of the second-order rate equation to the hydrodesulfurization process has been advocated because of its simplicity and use for extrapolating and interpolating hydrodesulfurization data over a wide variety of conditions. However, while the hydrodesulfurization process may appear to exhibit second-order kinetics at temperatures near to 395° C (745° F), at other temperatures the data (assuming second-order kinetics) does not give

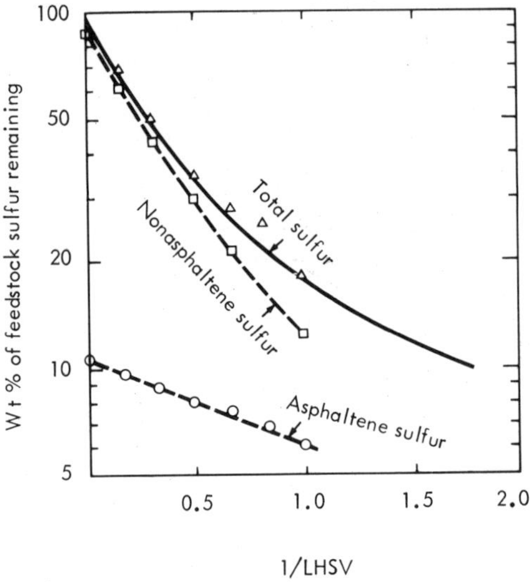

FIG. 3-5 Kinetics of the hydrodesulfurization of an Arabian light atmospheric residuum. [From J. W. Scott and A. G. Bridge, Origin and refining of petroleum, Advances in Chemistry Series, No. 103 (H. G. McGrath and M. E. Charles, eds.), American Chemical Society, Washington, D.C., 1971, p. 119.]

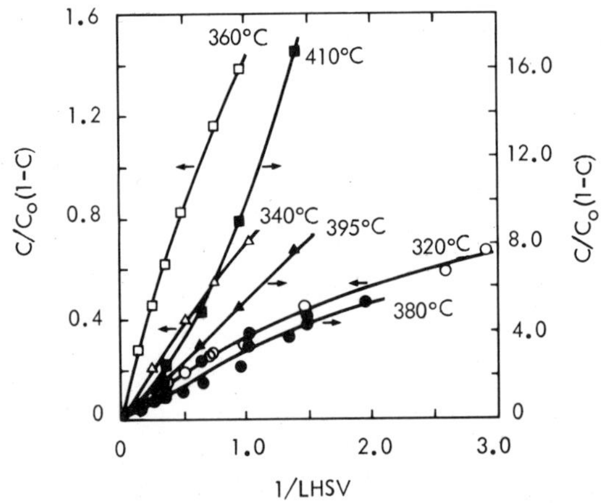

FIG. 3-6 Hydrodesulfurization of an atmospheric residuum (Kuwait) plotted as a second-order reaction. (From H. Ozaki, Y. Satomi and T. Hisamitsu, Proc. 9th World Petroleum Congress, 1963, Vol. 6, 97; reproduced with permission.)

a linear relationship (Fig. 3-6). On this basis, the use of two simultaneous first-order equations may be more appropriate insofar as the complexity of the sulfur compounds tends to increase with an increase in boiling point and the reactivity tends to decrease with complexity of the sulfur compounds, and residua (and, for that matter, the majority of heavy oils) may be expected to show substantial proportions of difficult-to-desulfurize sulfur compounds. It is anticipated that such an approach would be more consistent with the relative reactivities of various sulfur compound types observed for model compounds and for the various petroleum fractions that have been investigated.

Other kinetic work has shown that, for a fixed level of sulfur removal, the order of a reaction at constant temperature can be defined with respect to pressure:

$$k = \frac{1}{\text{LHSV}} (P_h)^n$$

where

P_h = hydrogen partial pressure

LHSV = liquid volume hourly space velocity

k = constant

n = order of the reaction

It has been concluded, on the basis of this equation, that the hydrodesulfurization of a Middle East residuum was first-order with respect to pressure over the range 800 to 2300 psi, although it does appear that the response to pressure diminishes markedly (and may even be minimal) above 1000 psi.

One marked effect of a hydrodesulfurization process is the buildup of hydrogen sulfide and the continued presence of this reaction product in the reactor reduces the rate of hydrodesulfurization. Thus, using the two first-order models, the effect of hydrogen sulfide on the process can be represented as:

$$\frac{k}{k_0} = \frac{1}{1 + k_1 \, P_{H_2S}}$$

where

k = rate constant with hydrogen sulfide present

k_0 = rate constant in the absence of hydrogen sulfide

k_1 = constant

Data obtained using this equation showed that a change in hydrogen sulfide concentration from 1 to 12% (by volume) could reduce by 50% the rate constants for the "easy-to-desulfurize" and the "difficult-to-desulfurize" reactions.

On the basis of the data available from kinetic investigations, the kinetics of residuum hydrodesulfurization may be represented by the following general equation:

$$\frac{-dS}{dt} = \left[\frac{P_H^n}{(1 + k_a A = k_s P_{H_2S})^m}\right] \sum_{i=1}^{i=j} k_i S_i$$

where

S = weight fraction of sulfur in the liquid phase

t = residence time

P_H = partial pressure of hydrogen

A = weight fraction of asphaltenes in the liquid phase

P_{H_2S} = partial pressure of hydrogen sulfide

S_i = weight fraction of sulfur associated with component i in the range of 1 to j.

k_a = adsorption constant for asphaltenes

k_s = adsorption constant for hydrogen sulfide

k_i = specific reaction rate constant for component i

The latter constant k_i in the above relationship is a function of the chemistry of the component, the catalyst activity, and the reaction temperature. Therefore:

$$k_i = k_0 \frac{A}{A_o} e^{-\Delta E/RT}$$

where

k_0 = reaction rate constant at standard catalyst activity

A_o = standard catalyst activity

A = catalyst activity

ΔE = activation energy

R = gas constant

T = absolute temperature

giving activation energies for the hydrodesulfurization of various residua in the range 27 to 35 kcal g-mol^{-1}. In this context, it was interesting to note that the deasphalting of Khafji residuum had no effect on the activation energy of 30 kcal g-mol^{-1} and it was suggested that the activation energies of the various components in a particular residuum may be approximately the same.

3-5 MACROMOLECULAR CONCEPTS OF HYDRODESULFURIZATION

In a mixture as complex as a residuum or heavy oil, the reaction processes can only be generalized because of difficulties in analyzing not only the products but also the feedstock as well as the intricate and complex nature of the molecules that make up the feedstock. Some indication has been given earlier in this chapter of the compound types that may constitute the sulfur-containing molecules in a residuum or heavy crude oil as well as the products that these molecular types produce as a result of hydrodesulfurization reaction. Although the deficiencies of current analytical techniques dictate that the actual mechanism of desulfurization remain largely speculative, some attempt has been made to determine the macromolecular chemical concepts that are involved in the hydrodesulfurization of heavy oils and residua.

Nevertheless, attempts have been made to determine the macromolecular changes that occur as a result of hydrodesulfurization by investigation of the changes that occur in the broad fractions, i.e., the asphaltenes, the resins, and the oils (see Chapter 1) that make up a residuum of heavy oil. Most of the reported work has focused attention on the asphaltene fraction because of the undesirable effects (e.g., coke deposition on the catalyst as well as the relatively high sulfur, nitrogen, and oxygen content) that this particular fraction (which is found in substantial proportions in residua and heavy oils) has on refining processes. However, there are reports which describe investigations into the overall effects of hydrodesulfurization on residua and heavy oils and which illustrate the macromolecular changes that occur during the process.

Thus, with respect to a Safaniya residuum, mild desulfurization appears to cause some loss, or denaturization, of the asphaltene fraction which was reduced in proportion from 17% by weight of the residuum to about 10% by

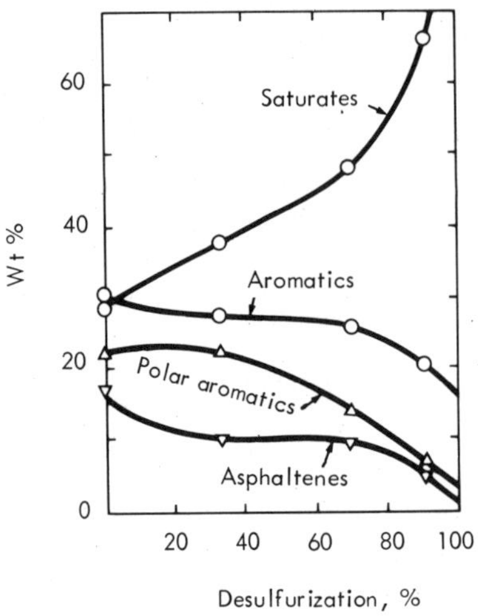

FIG. 3-7 Product composition from a Safaniya residuum in relation to reaction progress. [From H. V. Drushel, Preprints, American Chemical Society Division of Petroleum Chemistry, 1972, 17(4): F92.]

weight of the product. More severe conditions were required to reduce the proportion of asphaltenes in the product below 10% by weight and, even then, this could only be achieved by removing more than 70% of the sulfur that was originally present in the residuum. The disappearance of the asphaltenes under mild desulfurization conditions appeared initially to bring about an increase in the resin-type components (also referred to at times as polar aromatics) which then decreased in proportion under more severe desulfurization conditions. It is presumed that as the resin-type components were progressively desulfurized, aromatic hydrocarbons were the products, and hydrogenation of the less complex aromatic hydrocarbons yielded saturated materials (Fig. 3-7). It is interesting to note that sulfur appeared to be removed from the resin-type components (polar aromatics) and from the aromatics fraction at approximately the same rate (Fig. 3-8) while the percent sulfur in the asphaltenes precipitated from the products was about the same as the percent sulfur precipitated from the untreated original residuum (Fig. 3-8). In addition, the asphaltenes from the products contained a higher proportion of the total sulfur in the products than the asphaltenes in the original residuum relative to the total sulfur in the original residuum (Fig. 3-9).

Similar variations in product composition have also been noted for the thermal and catalytic hydrodesulfurization of a naturally occurring (Athabasca) bitumen (Fig. 3-10).

Insofar as the data presented above relate to two specific examples, it would be presumptuous to claim that it represented, in entirety, the means by which all residua and heavy oils are desulfurized. Nevertheless, it has been suggested by several studies that the overall pathway by which hydrodesulfurization of heavy oils and residua occurs involves a stepwise mechanism:

Asphaltenes ⟶ { polar aromatics / resin-type components } ⟶ aromatics* ⟶ saturates*

*The saturates and the aromatics fraction together are often referred to as the oils fraction (see Chapter 1).

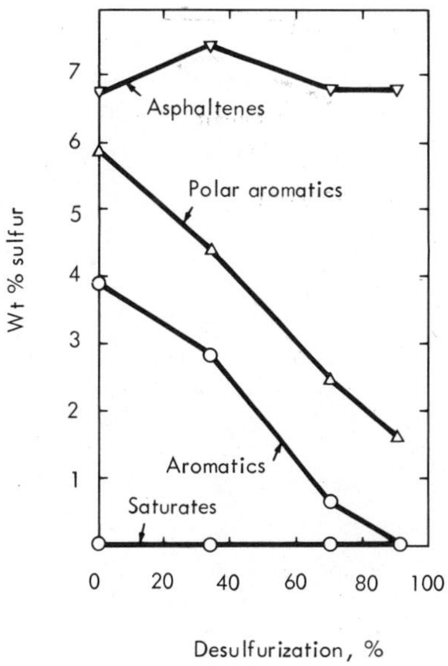

FIG. 3-8 Sulfur content of the product fractions from a Safaniya residuum in relation to reaction progress. [From H. V. Drushel, Preprints, American Chemical Society Division of Petroleum Chemistry, 1972, 17(4):F92.]

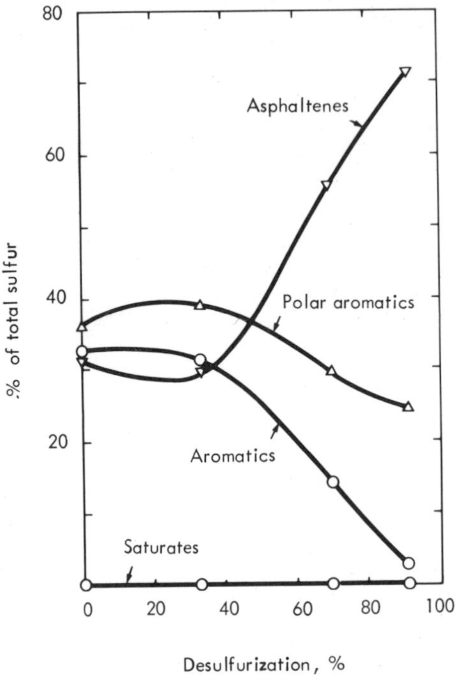

FIG. 3-9 Sulfur distribution in the product fractions from a Safaniya residuum in relation to reaction progress. [From H. V. Drushel, Preprints, American Chemical Society Division of Petroleum Chemistry, 1972, 17(4):F92.]

and that a direct step from either the asphaltenes or the resin-type components to the saturates is not considered a predominant pathway for hydrodesulfurization. There are also indications that the majority of the hydrodesulfurization occurs in the conversion of the resin-type components to the aromatics.

Because of their high molecular weight and complexity, the asphaltenes remain an unknown entity in the hydrodesulfurization process. There are indications that, with respect to some residua and heavy oils, removal of the asphaltenes prior to the hydrodesulfurization step brings out a severalfold increase in the rate of hydrodesulfurization and that, with these particular residua (or heavy oils), the asphaltenes must actually inhibit hydrodesulfurization. As a result of the "peculiar" behavior of the asphaltenes there have been several attempts to focus attention on the asphaltenes during hydrodesulfurization studies, while the other fractions of a residuum or

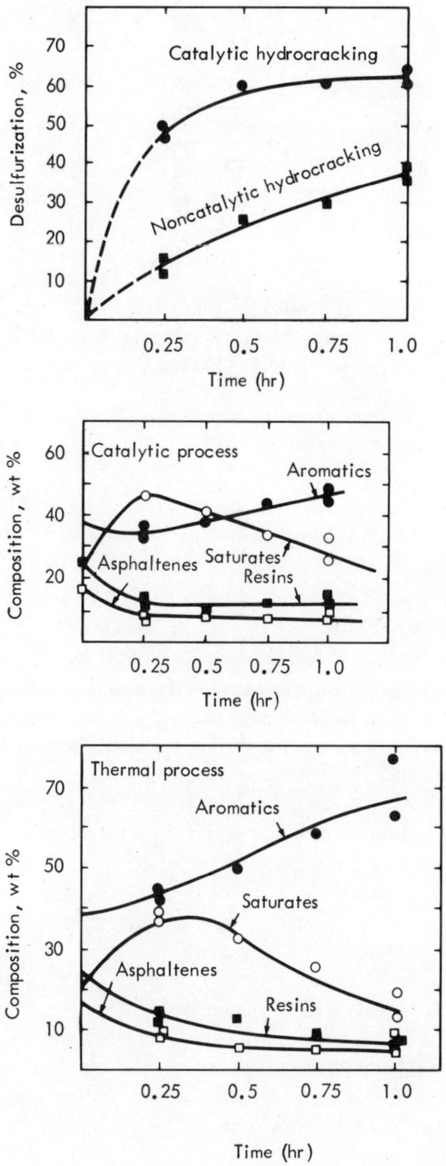

FIG. 3-10 Variations in product composition for the thermal and catalytic hydrodesulfurization of Athabasca bitumen. [From J. G. Speight and S. E. Moschopedis, The production of low-sulfur liquids and coke from Athabasca bitumen, Fuel Processing Technology, 1979, 2, 295.]

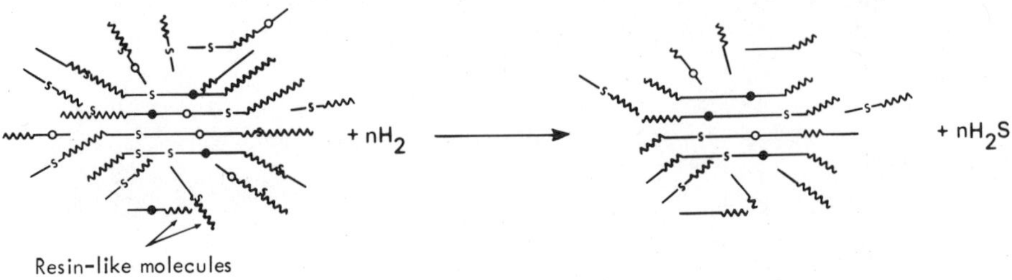

FIG. 3-11 Hypothetical mechanism for the hydrodesulfurization of asphaltenes and resins. Key: S sulfur; O vanadium; ● nickel; —aromatic rings; ∿∿∿ naphthenic rings. (From Proc. 6th World Petroleum Congress, 1963, Vol. 3, p. 303.)

heavy oil (i.e., the resins and the oils) have, on the other hand, been considered as higher-molecular-weight extensions of heavy gas oils about which desulfurization is fairly well understood. Thus, it is not surprising that several attempts have been made to represent the hydrodesulfurization process (with special emphasis on the asphaltene fraction) by structural macromolecular formulas. This, in itself, is contradictory since the the molecular formulas of the asphaltenes remain almost unknown and there may only be limited benefit (if any at all) in performing such an exercise.

Nevertheless, the asphaltenes have been considered to be micellar clusters with a tendency for the individual members to stack and align in a planar fashion not unlike graphite (Fig. 3-11) by means of interlayer forces with the resin-type components attached by intermolecular forces to the edge of the micelle. At high temperatures, the micelle is probably disrupted with the individual components behaving as a random mix of asphaltenes, resins, and oils. Under these conditions, conversion of the asphaltenes occurs when a fragment is lost through hydrodesulfurization (or even through hydrocracking). If the fragment or fragments are sufficiently large or of the nature that the remaining molecule cannot be precipitated by pentane, an asphaltene molecule has been converted to a resin molecule, but it appears that the loss of sulfur alone is not usually sufficient to change the tendency for asphaltenes to precipitate, and any of the largely speculative asphaltene structures that indicate that asphaltenes are predominatnly organosulfur polymers must certainly be suspect.

BIBLIOGRAPHY

Arey, W. F., Jr., Blackwell, N. E., III, and Reichle, A. D.: Advances in desulphurisation of residual fractions and asphalts, Proc. Seventh World Petrol. Cong., 4:167, 1968.

Beuther, H., and Schmid, B. K.: Reaction mechanisms and rates in residue hydrodesulphurisation, Proc. Sixth World Petrol. Cong., 3:297, 1963.

Crynes, Billy L. (ed.): Chemical Reactions as a Means of Separation—Sulphur Removal, Marcel Dekker, New York, 1977.

Drushel, Harry V.: Analytical characteization of residua and hydrotreated products, Preprints, Amer. Chem. Soc., Div. Petrol. Chem., 17(4):F92, 1972.

Frye, C. G., and Mosby, J. F.: Kinetics of htdrodesulphurisation, Chem. Eng. Prog., 63(9):66, 1967.

Gary, James H., and Handwerk, Glenn E.: Petroleum Refining—Technology and Economics, Marcel Dekker, New York, 1975.

McKetta, John J. (ed.): Advances in Petroleum Chemistry and Refining, Vol. 3, Interscience, New York, 1960.

McKinley, Joseph B., The hydrodesulphurisation of liquid petroleum fractions, Catalysis, 5:405, 1957.

Richardson, Ryden L., and Alley, Starling K.: Consideration of catalyst pore structure and asphaltenic sulphur in the desulphurisation of resids, Hydrocracking and Hydrotreating, ACS Symposium Series 20:136, 1975.

Schuman, S. C., and Shalit, H.: Hydrodesulfurisation, Catalysis Reviews, 4:245, 1970.

Scott, J. W., and Bridge, A. G.: The continuing development of hydrocracking, Origin and Refining of Petroleum, ACS Advances in Chemistry Series No. 103:113, 1971.

Speight, James G.: The Chemistry and Technology of Petroleum, Marcel Dekker, New York, 1980.

Speight, James G.: The Structure of Petroleum Asphaltenes—Current Concepts, Information Series No. 81, Alberta Research Council, 1978.

Speight, James G., and Moschopedis, Speros E.: The production of low-sulphur liquids and coke from athabasca bitumen, Fuel Processing Technology, 2:295, 1979.

Speight, James G., and Moschopedis, Speros E.: Some observations on the molecular nature of petroleum asphaltenes, Preprints, American Chemical Society, Division of Petroleum Chemistry, 24(4):910, 1979.

4

DESULFURIZATION DURING REFINERY OPERATIONS

The depletion of the reserves of conventional petroleums has focused the attention of refiners on secondary feedstocks (residua) and the heavier asphaltic oils. However, the move to process these less desirable feedstocks has led to legislation to control the amount of sulfur in the various grades of liquid fuels that are produced which, in turn, has meant that processing facilities must be adjusted accordingly to effectively remove the sulfur from the feedstocks (or products) as an integral part of the processing sequence.

The need to remove sulfur compounds from petroleum arises not only from the objectionable odor which the thiols impart to petroleum products but also from the instability which sulfur compounds appear to promote in petroleum products. For example, free sulfur, which can be formed in a product by the oxidation of hydrogen sulfide:

$$2H_2S + O_2 \rightarrow 2H_2O + 2S$$

is noted for the complex series of reactions that it will undergo with a variety of hydrocarbons of the type that might be found in, say, fuel oil products. On the other hand, thiophenes, aliphatic mercaptans, and general organic sulfides appear to have little effect, whereas disulfides and polysulfides ($R-S_n-R$, where n = 2 or more) actively promote the formation of sludges. However, even though a particular class of sulfur compounds may appear to have very little deleterious effect on the product in which it is found, it is actually the reactions which this material can promote or the products which it can form or even the ultimate use of the petroleum fraction that dictates whether or not (and to what extent) the sulfur compounds should be removed. From the general aspects of petroleum refining, it must be presumed that sulfur compounds, no matter what their chemical constitution, should be removed to preserve product quality and to protect refinery equipment.

TABLE 4-1 Sulfur Concentration in Petroleum Cokes from Various Feedstocks

Feedstock	API Gravity	Sulfur (wt %)	Sulfur in Coke (wt %)	S% (Coke) / S% (Feedstock)
Elk Basin, Wyo., residuum	2.5	3.5	6.5	1.83
Hawkins, Tex., residuum	4.5	4.5	7.0	1.55
Kuwait, residuum	6.0	5.37	10.8	2.01
Athabasca (Canada), bitumen	7.3	5.3	5.65	1.06
West Texas, residuum	–	3.5	3.06	0.875
Boscan (Venezuela), crude oil	10.0	5.0	5.0	1.0
East Texas, residuum	10.5	1.26	2.57	2.04
Texas Panhandle, residuum	18.9	0.6	0.6	1.00

Sulfur removal, as practiced in various refineries, can take several forms such as concentration in refinery products (e.g., coke; Table 4-1), hydrodesulfurization, or chemical removal (acid treating, caustic treating—i.e., sweetening or finishing processes). Nevertheless, the desulfurization of petroleum is almost universally accomplished by the catalytic reaction of hydrogen with the sulfur-containing constituents of petroleum to produce hydrogen sulfide which can be readily separated from the liquid products. However, there are certain other refinery processes which are adaptable to residua and heavy oils and which may be effective for reducing—but not necessarily effective for removing completely—the sulfur content of these feedstocks, and in a text of this nature these processes should be given some consideration and are, therefore, described briefly in the following pages.

4-1 DISTILLATION

Distillation was the first method by which petroleum was refined, but this was not so much a refining method as a method for separation of the crude oil into fractions, such as kerosene, for which there was a demand and hence a ready market. Nevertheless, as petroleum refineries have evolved into the present-day complexes, distillation units may still find a place, depending upon the characteristics of the crude oil feedstock, in a refinery sequence. The separation of the crude oil into fractions is accomplished in two stages: in the first stage, the feedstock is fractionated at atmospheric pressure and then, in the second stage, the atmospheric residuum is fractionated in a unit operated under a high vacuum (Fig. 4-1). Of all the units in a refinery, the distillation section is perhaps the most versatile as it is necessary for this unit to have the greatest flexibility in terms of adjusting to feedstocks varying markedly in quality as well as producing a wide range of products.

Distillation is not a desulfurization process insofar as the sulfur in the feedstock is not completely (if at all) eliminated. Maximum permissible temperatures to which the feedstock is heated are of the order of 350° C (660° F) which essentially minimizes the tendency for thermal decomposition of the components of the feedstock to occur. Hence, distillation, insofar as the sulfur-containing constituents of crude oils are concerned, is actually a concentration process by which the majority of the sulfur-containing compounds (which are usually of high molecular weight) are concentrated into the higher-boiling fractions (Table 4-2; see also Figs. 2-2, 2-3, 2-4, and 2-5).

Distillation is also a means by which the character, especially the sulfur content, of a residuum may be adjusted. For example, inspections

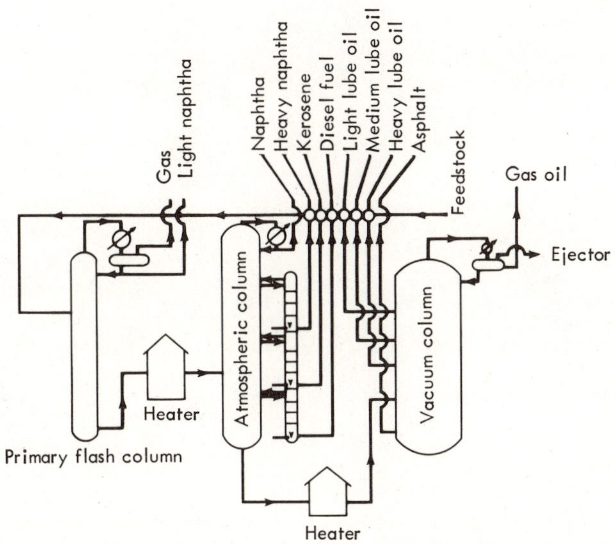

FIGURE 4-1 Schematic representation of a refinery distillation sequence.

TABLE 4-2 Distillation Products from an Arabian Crude Oil

	Gasoline	Kerosene	Gas Oil	Residuum
Yield, wt %	15	12	17	55
Range, °C	<150	150–230	230–340	340+
Specific gravity	0.70	0.78	0.85	0.97
API gravity	70.6	49.9	35.0	14.3
Sulfur, wt %	0.02	0.2	1.4	3.9

TABLE 4-3 Inspections of Various Residua to Illustrate the Increase in Sulfur Content from Atmospheric to Vacuum Residua

	Arabian Light	Arabian Heavy	Alaska (Sag River)
Atmospheric residuum (650° F+):			
Yield on crude, vol %	43.0	52.8	53.0
Gravity, °API	16.2	11.1	15.5
Sulfur, wt %	3.1	4.5	1.6
Metals (Ni + V), ppm	33	124	46
Vacuum residuum (1050° F+):			
Yield on crude, vol %	13.2	26.0	18.3
Gravity, °API	6.5	4.0	7.6
Sulfur, wt %	4.1	5.7	2.3
Metals (Ni + V), ppm	100	239	126
Carbon residue (Ramsbottom)	22	24	18
Asphaltenes (pentane insolubles), wt %	12	25	10

of various crude oil residua (Table 4-3) show that, for any particular crude oil, the vacuum residuum is virtually always higher in sulfur than an atmospheric residuum from the same crude oil. Thus, although distillation is the usual primary means by which a crude oil is processed, it may be completely bypassed in the case of an extremely heavy crude oil in favor of whole-crude processing by any of the more suitable thermal methods. Indeed, with respect to heavy oils, distillation as such may never be applied but may be used as a flash technique in which the whole crude is introduced into a vessel and the temperature is such that all of the volatile material boiling below a predetermined temperature is removed in toto as an overhead oil without any attempt at fractionation. This total fraction from the heavy feedstock may be subdivided at a later stage of the refinery operation or serve as a feedstock (without fractionation) for a cracking unit. In short, distillation is, at best, looked upon as a means by which the lower-boiling fractions can be separated from a feedstock prior to being subjected to a suitable conversion (or refining) method. It is, in fact, the means by which the undesirable higher molecular weight materials are removed from the feedstock as atmospheric or vacuum residua. It would, indeed, be a very rare occasion if the distillation process actually served as an efficient means of desulfurization rather than a concentration process.

4-2 VISBREAKING

Visbreaking is a relatively mild thermal cracking process that is used to reduce the viscosity of heavy straight-run residua. The heavy residua are often blended with lighter heating oils to produce fuel oils of acceptable viscosity which may be considered to be a waste of the higher-value light heating oil. Thus, reducing the viscosity of the heavy residuum by this particular process, the amount of light heating oil required for the blending operation is markedly reduced and more favorable economics result.

Visbreaking conditions range from 450° C (840° F) to about 510° C (950° F) at the heating coil outlet with pressures varying from 50 to 300 psi. Cracking is initiated in the liquid phase and continues in the reaction chamber (Fig. 4-2) at slightly reduced temperatures (445 to 455° C; 830 to 850° F) but

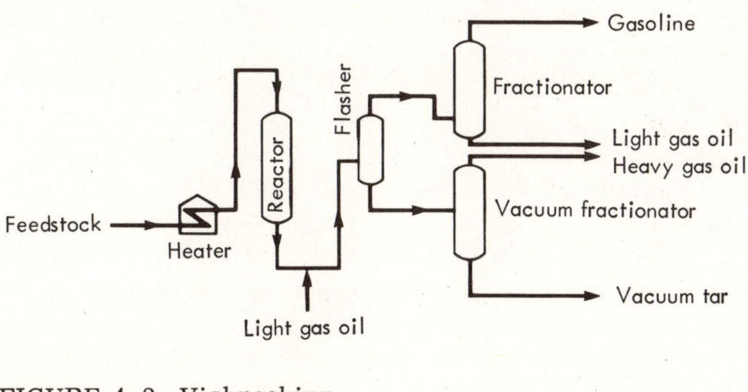

FIGURE 4-2 Visbreaking.

TABLE 4-4 Visbreaking of a United States (Louisiana) Residuum

Feedstock			
Specific gravity		0.987	
API gravity		11.9	
Conradson carbon, wt %		10.6	
Viscosity SFS (210° F)		495	
Sulfur, wt %		0.61	
Product yields, vol %			
$\leq C_4$, vol %		1.6	
C_5 to 220° C Naphtha, vol %		6.2	
220–340° C Gas oil, vol %		6.3	
>340° C (Residuum), vol %		88.4	
	Naphtha	Gas Oil	Residuum
Product quality			
Specific gravity	0.748	0.851	0.990
API gravity	57.7	34.8	11.4
Conradson carbon, wt %	–	0.01	15.0
Viscosity (SFS, 210° F)	–	–	160
Sulfur, wt %	0.26	0.33	0.58

the reactions are not allowed to proceed to completion; the hot liquid reaction mix is quenched by the addition of a light gas oil. Cracked products from the reaction chamber are then fed to a flasher followed by fractionation into various streams.

Visbreaking is usually claimed to have little, or no, effect on the sulfur compounds in a residuum as the sole purpose of the process is to reduce the viscosity of the feedstock. However, it is capable of producing product streams that have reduced sulfur contents (relative to the feedstock; Table 4-4) and which are more amenable to subsequent desulfurization; these desulfurization stocks can then be blended back with the original feedstock or with the residuum (from the visbreaking) to produce an overall

TABLE 4-5 Visbreaking of a Heavy Canadian (Athabasca) Natural Bitumen

Feedstock			
Gravity, °API	8.6		
Distillation	Vol %	°F	
	IPB	–	
	5	430	
	10	560	
	30	820	
	50	1010	
Sulfur	4.8		
Nitrogen	0.4		
Conradson carbon, wt %	13.5		
Product yields (wt % on whole bitumen)			
$\leqslant C_4$	3		
C_5/400° F	7		
400/650° F	21		
650/1000° F	35		
Residue	34		
Total	100		
Product qualities	C_5/380 (195° C)	380/650 (195/345° C)	650/1000 (345/540° C)
Gravity, °API	54	25.8	13.2
Sulfur, wt %	2.02	2.11	3.91
Nitrogen, ppm	110	450	2800
Bromine number	115	38	20
Conradson carbon, wt %	–	–	0.23

low-sulfur stream. Visbreaking has been seriously considered as a potential primary process for the upgrading of the heavy bituminous material that can be separated from the Athabasca oil sand (Table 4-5).

4-3 CATALYTIC CRACKING

Catalytic cracking is a process by which high-boiling hydrocarbon stocks can be converted into gasoline fractions. Catalytic cracking processes can generally be classified as a moving bed process (e.g., the Thermofor catalytic cracking process—Fig. 4-3) or as a fluid bed process (e.g., fluid catalytic cracking—Fig. 4-4). The catalysts employed for both types of processes are very similar chemically but do differ physically—moving bed units often employ catalysts in the form of beads or cylinders (1/8 to 1/4 in. diameter) while fluid bed units usually employ the catalyst in much smaller form where particle sizes may be of the order of 50 μm (50×10^{-4} cm); catalysts are usually modified hydrated alumina silicates.

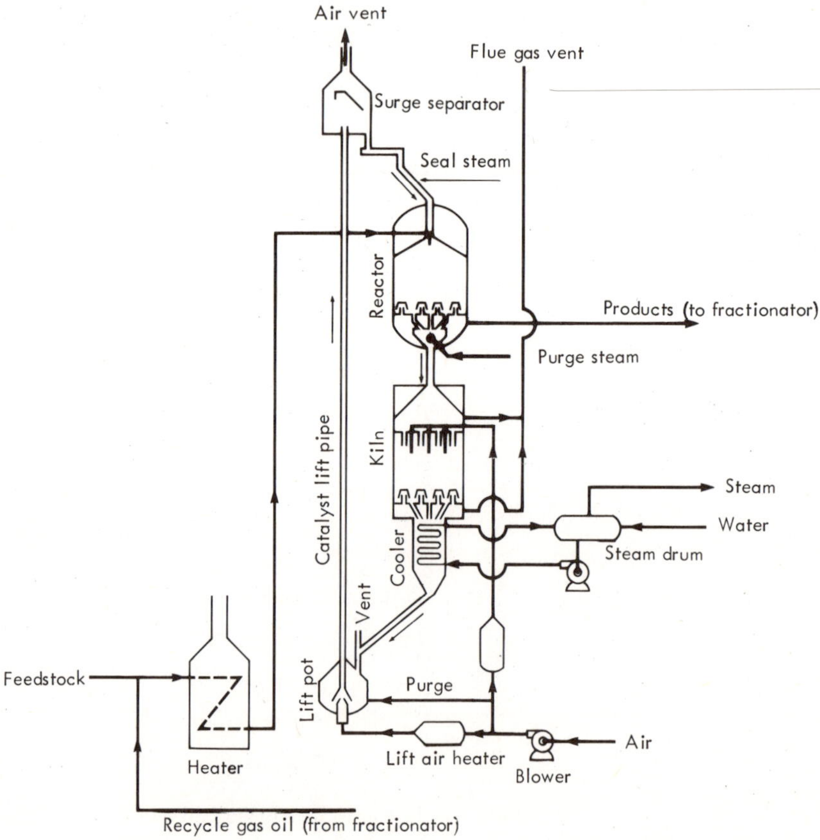

FIGURE 4-3 The Thermofar catalytic cracking unit.

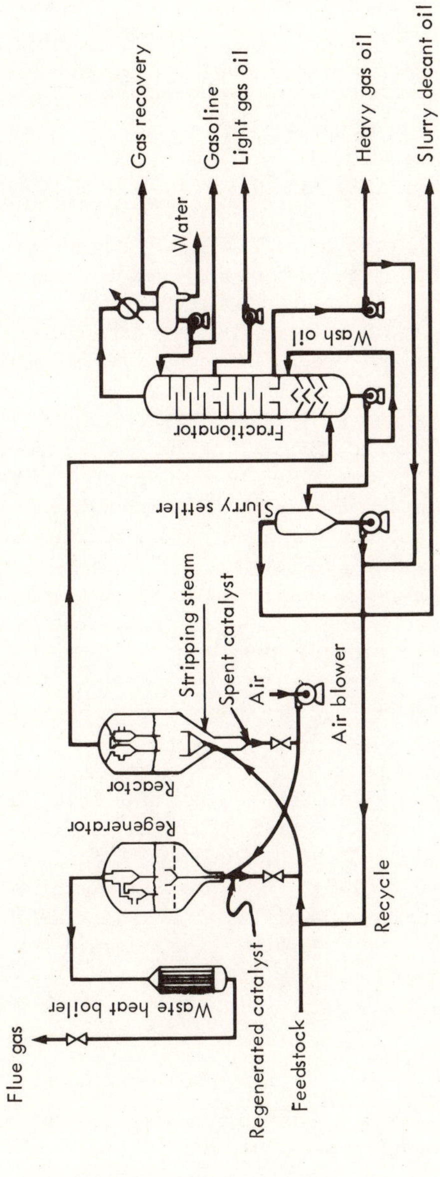

FIGURE 4-4 A fluid catalytic cracking unit.

Catalytic cracking processes produce coke and overheads (volatile materials). The coke deposits on the catalyst and, therefore, markedly decreases catalyst activity. This problem is surmounted by transference of the catalyst from the reactor to a vessel (regenerator) where the coke is burned from the catalyst in air; movement of the catalyst between the vessels is continuous, thereby maintaining feedstock flow and production of products. Reactor temperatures are usually of the order of 465 to 510°C (870 to 950°F) and 10 to 16 psi while regenerator temperatures may be substantially higher at 565 to 675°C (1050 to 1250°F) and 8 to 10 psi; the feedstock temperatures vary from 315 to 455°C (600 to 850°F) at point of entry to the reactor.

Although catalytic cracking was originally designed to convert the heavier, more refractory gas oils to gasoline, process modifications have allowed the feedstock types to include the heavy oils and residua. Naturally, the way in which a given feedstock will crack depends on a number of process variables such as feedstock composition, boiling range and hydrocarbon types present, reactor conditions, catalyst type, etc., and it is not surprising that the heavier feedstocks present special problems when used to feed a catalytic cracker. The very nature of the process dictates the use of a catalyst and the susceptibility of the catalyst to the various nitrogen compounds, as well as the metals (iron, nickel, vanadium, and copper) in the feedstock, represents a major drawback to using heavy oils and residua as catalytic cracker feedstock. For example, if the metals concentration (expressed in ppm) on the catalyst fulfills the following relation:

$$4V + 14Ni + Fe + Cu > 1000$$

then the catalyst is severely contaminated and may actually have to be re-replaced.

Nevertheless, it is possible to process the heavier feedstocks in catalytic crackers and bring about some degree of desulfurization in the process. The effectiveness, and degree, of desulfurization depends upon the amount of sulfur in the feedstock (Tables 4-6 and 4-7) and it is generally believed that to optimize use of a catalytic cracking unit, feedstocks should be treated to remove excess asphaltic material and metals by processes such as visbreaking, coking, or deasphalting (see elsewhere in this chapter; Fig. 4-5) thereby prolonging catalyst activity.

4-4 COKING

Coking is generally regarded as a relatively severe thermal cracking process designed to maximize the conversion of petroleum residua and other heavy oils to gas and liquid products. Often the gas oil fraction is the major product and is used primarily as a feedstock for catalytic

TABLE 4-6 Fluid Catalytic Cracking of Residua from the Hydrodesulfurization of Alaskan Crude Oils

	Sag River 375° F+ (190° C+)	Put River 375° F+ (190° C+)
Residuum properties		
API gravity	21.5	21.1
Sulfur, wt %	0.1	0.3
Aniline point, ° F	199	192
Conradson carbon, wt %	2.2	4.8
Nitrogen, wt %	–	0.25
Conversion, vol %	80.2	78.0
Yields		
C_4, vol %	20.1	28.5
C_5-430° F gasoline, vol %	60.8	58.9
430-650° F, vol %	14.5	15.1
Decanted oil	5.3	6.9
Coke, wt %	7.8	8.1
Gasoline properties		
API gravity	57	56
Sulfur, wt %	0.01	0.03
Distillate properties		
API gravity	17.0	18.0
Sulfur, wt %	0.24	0.50
Catalyst makeup rate		
lb/bbl Fresh feed	0.5	1.4

cracking units. The coking process usually concentrates sulfur in the feedstock into the coke fraction and yields liquid products which are substantially lower in sulfur than the feedstock and which can be fairly readily desulfurized.

TABLE 4-7 Fluid Catalytic Cracking of Deasphalted Feedstocks

	Virgin	Hydrotreated
Feedstock		
Specific gravity	0.970	0.931
API gravity	14.3	20.5
Aniline point, °C	96	101.5
Sulfur, wt %	3.55	0.30
Nitrogen, wt %	0.20	0.11
Distillation, °C		
10% vol	530	480
30% vol	570	545
50% vol	590	580
Conversion, vol %	70	70
Product Yields		
$\leqslant C_4$, vol %	22.1	21.6
Gasoline, C_5-204°C (ASTM) EP, vol %	58.5	61.0
Light cycle oil, vol %	24.0	24.0
Decanted oil, vol %	6.0	6.0
Coke, wt %	7.0	4.5
Product Properties		
Gasoline, C_5-204°C		
Specific gravity	0.759	0.745
API gravity	54.9	58.4
Sulfur, wt %	0.39	0.02
Light cycle oil		
Specific gravity	0.959	0.929
API gravity	16.1	20.8

TABLE 4-7 (Continued)

	Virgin	Hydrotreated
Product Properties (continued)		
Light cycle oil (continued)		
Sulfur, wt %	3.9	0.2
Viscosity, cSt/210° F	0.8	0.7
Decanted oil		
Specific gravity	1.058	1.013
API gravity	2.2	8.2
Sulfur, wt %	9.7	0.49
Viscosity, cSt/210° F	21	16

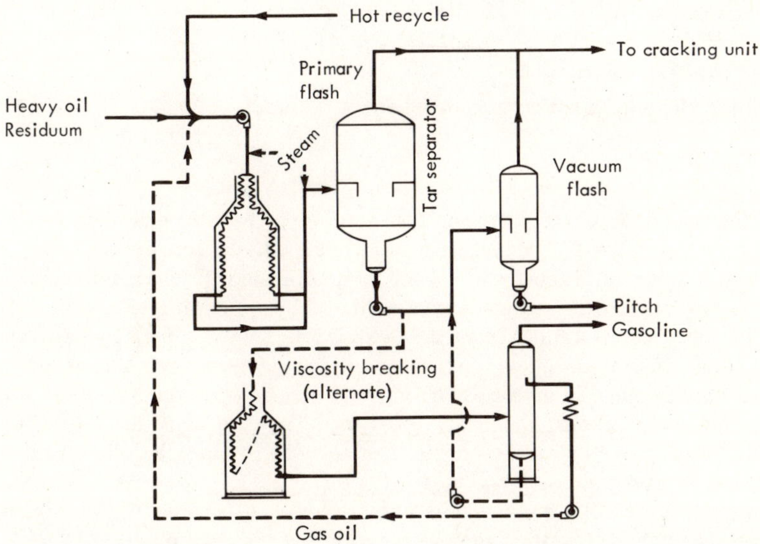

FIGURE 4-5 Various methods of feedstock preparation for a cracking unit.

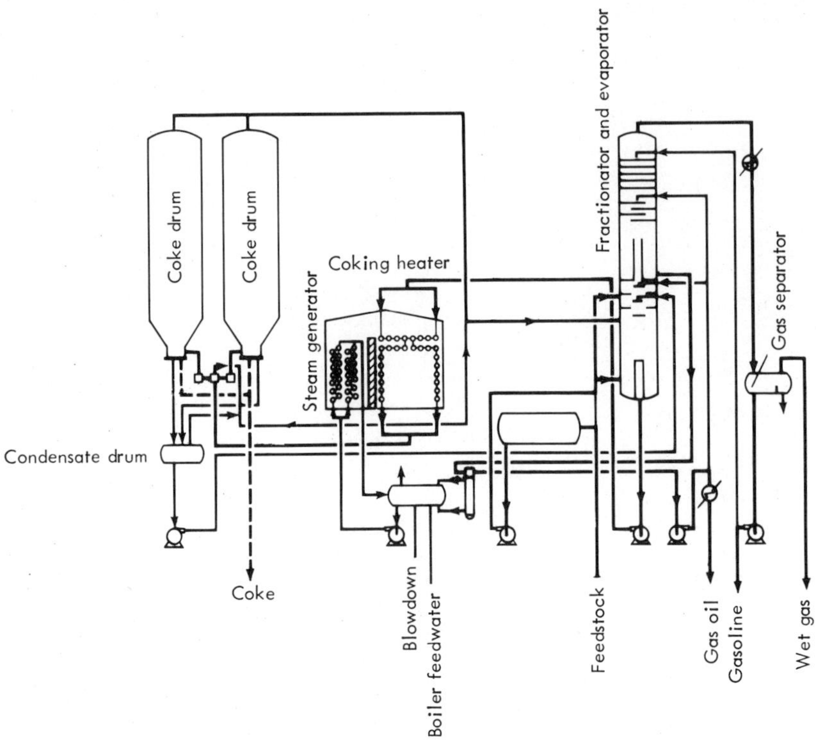

FIGURE 4-6 The delayed coking process; □ = switch valves.

Coking processes are different from the other thermal processes found in a refinery insofar as the reaction times may be longer and the reactions are usually allowed to proceed to completion (in contrast to, say, visbreaking where the reactions are terminated by quenching with a gas oil fraction). The coke obtained from the coking processes is usually used as a fuel for the process although marketing for specialty uses, such as electrode manufacture, increases the value of the coke. But, because of the tendency of the process to concentrate the feedstock sulfur in the coke the coker feedstock may have to be chosen carefully to produce a coke of sufficiently low sulfur content for a specialty use.

Two particular coking processes are currently used on a regular basis in the refining industry with a view to maximizing the conversion of heavy oils and residua; these are described, in turn, as follows.

A. Delayed Coking

Delayed coking is a semicontinuous process that can be applied to the conversion of most types of residua and heavy oils to cracking feedstock—gas oil may be the major product. In the process, heated feedstock is charged to a fractionating tower (Fig. 4-6) and the residuum from this tower is then reheated and passed to large coking drums where the reactions are allowed to proceed to completion. Volatile materials from the coke drum are conveyed to the fractionator while the drum is progressively filled with solids (coke).

Depending on the size of the refinery installation, the flow of the heated feedstock is alternated between two or more drums in order to allow coke removal from one drum (usually by mechanical or hydraulic methods) while another drum is onstream. The decomposition (coking) reactions are usually initiated at temperatures of the order of 480° C to 510° C (900 to 950° F) and are allowed to continue in the drum at 455° C to 480° C (850 to 900° F); reaction pressures are not usually greater than 100 psi.

The feedstocks that can be processed in this manner vary from light to heavy residua and also include heavy oils and even a heavy natural bitumen. In every case, the liquid products show substantial reductions in sulfur content relative to the feedstock (Table 4-8) with the overall tendency either to release the sulfur in the form of gaseous products (predominantly hydrogen sulfide) or to concentrate the majority of the feedstock sulfur in the coke (Table 4-9).

B. Fluid Coking

Fluid coking is a continuous process that employs the fluidized solids technique for the conversion of heavier low-grade feedstocks to lighter, more valuable, products. The feedstock is sprayed on to a fluidized bed of hot coke particles which is agitated by the gaseous products rising through the bed in the reactor. The fluidized solids technique permits the use of higher temperatures (than delayed coking) but without the usual overabundance of coke formation because of the shorter contact times with the result that higher yields of liquid products are produced.

The process employs two vessels (Fig. 4-7)—a reactor and a burner with coke particles circulated between the two to transfer heat (generated by burning excess coke in the burner) to the reactor. The preheated feedstock (260 to 370° C; 500 to 700° F) is injected directly onto the hot coke (480 to 565° C; 895 to 1050° F), at approximately atmospheric pressure, to crack to additional coke and liquid products which leave the reactor as overhead oil.

TABLE 4-8 Delayed Coking of Various Residua

Source	South Louisiana	Kuwait Virgin	Kuwait Hydrodesulfurized
Feedstock properties			
Gravity, °API	12.3	6.7	16.1
Sulfur, wt %	0.68	5.22	0.66
Carbon residue, wt %	13.0	19.8	9.1
Product yields			
C_4, wt %	7.8	7.6	9.0
C_5-400° F, vol %	22.8	26.7	22.0
400-650° F, vol %	18.4	28.0	41.9
Gas oil, vol %	37.6	18.4	19.1
Coke, wt %	23.7	30.2	18.5
Product properties			
Naphtha			
Gravity, °API	56.5	54.1	57.5
Sulfur, wt %	0.2	0.84	0.04
Furnace Oil			
Gravity, °API	34	28.5	37.0
Sulfur, wt %	0.3	3.15	0.18
Nitrogen, wt %	0.08	0.04	0.05
Gas Oil			
Gravity, °API	23.0	18.3	23.3
Sulfur, wt %	0.6	4.40	0.66
Nitrogen, wt %	0.13	0.15	0.19
Coke			
Sulfur, wt %	1.3	7.5	1.7

West Texas					
Virgin	Hydrode-sulfurized	Oklahoma	California	Gilsonite	Athabasca Bitumen
8.9	15.2	13.0	12.0	2.0	7.3
2.96	0.64	1.15	1.6	0.3	5.3
17.8	9.3	14.1	9.4	26.0	17.9
10.9	9.5	9.8	14.9	17.6	8.2
28.9	20.1	20.4	22.5	45.2	20.3
16.5	31.7		36.5	8.5	58.8
26.4	27.5	57.2	16.7		
28.4	20.7	23.6	19.1	45.0	21.0
53.7	57.1	56.4	57.9	57.2	51.9
0.6	0.09	–	0.9	–	1.9
33.9	35.3	–	27.5	–	–
1.08	0.19	–	–	–	2.7
0.06	0.08	–	–	–	–
21.6	23.8	28.0	14.9	15.9	–
1.93	0.45	–	1.1	–	3.8
0.25	0.26	–	–	–	–
4.5	1.6	–	–	–	5-10

TABLE 4-9 Sulfur (and Nitrogen) Distribution in the Products from the Delayed Coking of Various United States Residua

Product	Sulfur (%)	Nitrogen (%)
Gas	30	–
Naphtha	5	1
Kerosene	} 35	2
Gas oil		22 } 24
Coke	30	75
Totals:	100	100

Source: Reproduced with permission from J. H. Gary and G. E. Handwerk, Petroleum Refining: Technology and Economics, 1975, courtesy Marcel Dekker, Inc.

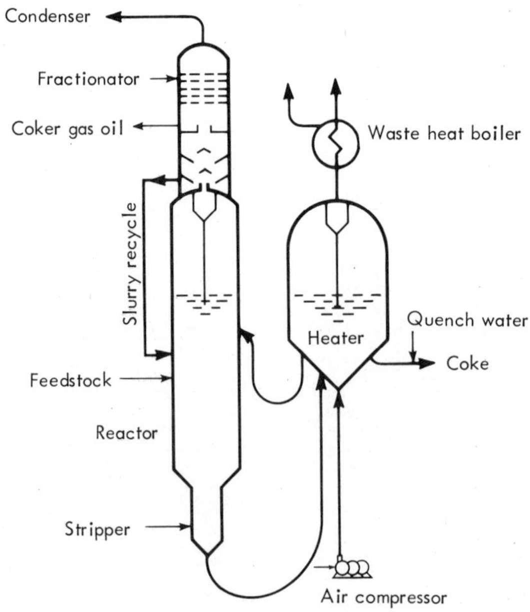

FIGURE 4-7 A fluid coking unit.

The feedstocks charged to a fluid coker may be any type of petroleum residue or heavy oil where the carbon residue falls into the range 5 to 50 wt % or those materials having an API gravity less than 20 (Tables 1-1 and 2-2). The bed temperature in the burner is of the order of 590 to 650° C (1095 to 1200° F), and any excess coke that is not removed as part of the burning is periodically removed from the burner. The coke yield from the process may be as little as 1.2 or as high as 1.7 times the Conradson residue of the feedstock. As with delayed coking, the fluid coking process is capable of producing liquid products with substantially lower sulfur contents than the feedstock (Table 4-10), but part of the sulfur in the feedstock is concentrated in the coke. There is some elimination of sulfur into the gaseous products but uses for the coke depend very much upon the amount of feedstock sulfur.

4-5 HYDROCRACKING

Hydrocracking is probably the most versatile of petroleum refining processes because of its applicability to a wide range of feedstocks. Chemically, hydrocracking can be regarded as a combination of cracking, hydrogenation, and isomerization. It is also a treating operation since elements such as sulfur, nitrogen, and oxygen are almost completely eliminated during the process (see Chapter 3). The process employs a catalyst and an environment of hydrogen at pressures of the order of 800 to 2500 psi.

The reactions that occur during the hydrocracking process require the use of a dual-function catalyst which has the capability of promoting not only the cracking reactions but is also capable as serving as a hydrogenation catalyst. The acidic function of the catalyst is usually provided by the catalyst base, and materials such as acid-treated montmorillonite (as well as alumina, silica-alumina, and zeolites) are employed to support the hydrogenation function which may contain metals such as nickel, tungsten, platinum, or palladium. However, these types of catalyst tend to be sensitive to nitrogen in the feedstock and a purification stage prior to hydrocracking may be necessary. Various types of catalyst bed have been proposed for hydrocracking reactors but generally fall into the fixed-bed or ebullating-bed category.

The choice of processing schemes for a given hydrocracking application depends upon the nature of the feedstock as well as the product requirements. The process can be simply illustrated as a single-stage or as a two-stage operation (Fig. 4-8). The single-stage process can be used to produce gasoline but is more often used to produce middle distillate from heavy vacuum gas oils, while the two-stage process was developed primarily to produce high yields of gasoline from straight-run gas oils and the first stage may actually be a purification step to remove sulfur-containing (as well as nitrogen-containing) organic materials. Both processes use an

TABLE 4-10 Fluid Coking Various Feedstocks

	Los Angeles Basin			Texas Panhandle
	Vacuum Residuum	Deasphalter Residuum	Visbreaker Residuum	Vacuum Residuum
Feedstock properties				
Specific gravity	1.024	1.078	1.106	0.951
API gravity	6.7	–	–	17.3
Conradson carbon, wt %	17	33	41	11
Sulfur, wt %	2.1	2.3	2.1	0.7
Product yields				
C_4, vol %	11	13	15	9
C_5 to 221°C Naphtha, vol %	17	18	14	21
221-545°C gas oil, vol %	62	45	32	69
Coke, wt %	21	36	48	12
Product quality				
Naphtha (C_5 to 221°C)				
Specific gravity	0.759	0.763	0.759	0.743
API gravity	54.9	63.9	54.9	58.9
Sulfur, wt %	–	1.7	0.8	0.2
Gas oil (221-545°C)				
Specific gravity	0.953	0.973	0.966	0.882
API gravity	17.0	13.9	15.0	28.9
Sulfur, wt %	1.7	2.2	1.7	0.4

Kuwait	South Louisiana		Hawkins	Zaca
Vacuum Residuum	Vacuum Residuum	Atmospheric Residuum	Vacuum Residuum	Atmospheric Residuum
1.032	0.989	0.948	1.043	1.039
5.6	11.6	17.8	4.2	4.7
21.8	13	5	24.5	19
5.5	0.6	0.5	4.3	7.8
14	11	8	13	10.5
21	21	17	19.5	20.5
48	61	74	52.0	61
28	17	8	27.5	17.5
0.755	0.755	0.755	0.755	0.784
55.9	55.9	55.9	55.9	49
0.5	0.2	0.1	0.9	4.6
0.963	0.922	0.910	0.953	0.973
15.4	22.0	24.0	17.0	14
4.7	0.5	0.4	3.7	6.2

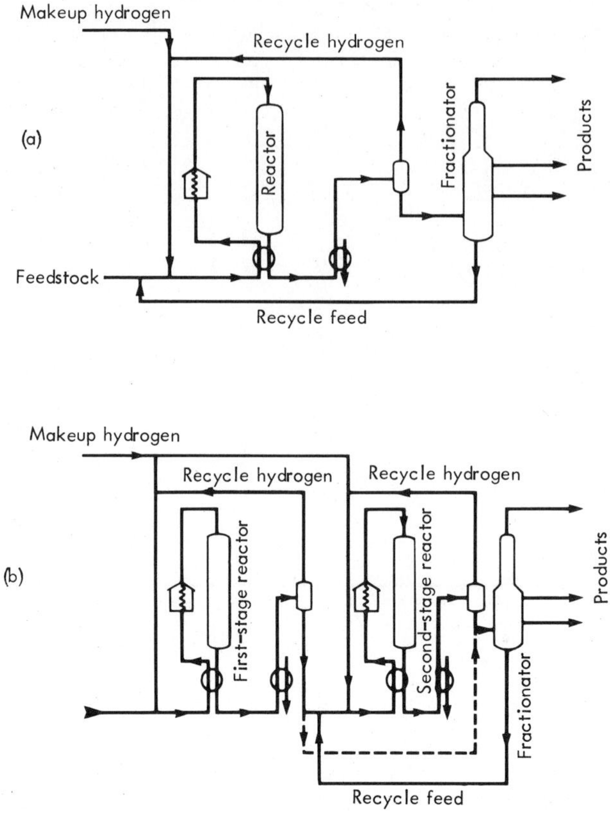

FIGURE 4-8 Schematic representation of (a) a single-stage and (b) a two-stage hydrocracking unit.

extinction recycle technique to maximize the yields of the desired product.

Product yields (Fig. 4-9) and hydrogen consumption (Fig. 4-10) vary with the feedstock, but the process can be utilized in many different ways. One of the major advantages of hydrocracking is that it may be used to process the higher-boiling refractory feedstocks which may be produced by catalytic cracking or by any of the coking processes. Hydrocracking can also be applied to the conversion of the more difficult feedstocks such as residua and heavy oils and there are a variety of hydrocracking processes that are designed specifically for this particular use. These processes (since they are also used as a means of hydrodesulfurization) will be dealt with detail in Chapter 7. It is worthy of note at this point that there is often very little effort (in fact, it is often difficult) to differentiate between a hydrocracking process and a hydrodesulfurization process; in actual practice, the reactions that are used to chemically define the processes (i.e., cracking and subsequent hydrogenation of the fragments, hydrogenation of unsaturated material, hydrodesulfurization, hydrodenitrogenation, etc.) may all occur (or be encouraged to occur) in any one particular process. Thus, hydrocracking of a feedstock will, in all likelihood, be accompanied by hydrodesulfurization, thereby producing not only low-boiling products but low-boiling products that are low in sulfur (Table 4-11).

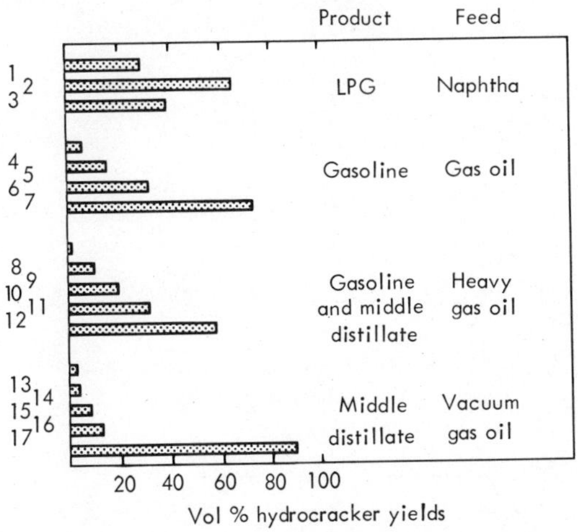

1: C_3 + gas
2: C_4
3: $C_5 - C_6$
4: C_3 + gas
5: C_4
6: $C_5 - C_6$
7: C_7 - 200°C (390°F)
8: C_3 + gas
9: C_4
10: $C_5 - C_6$
11: C_7 - 177°C (350°F)
12: 177 - 232°C (350 - 450°F)
13: C_3 + gas
14: C_4
15: $C_5 - C_6$
16: C_7 - 149°C (300°F)
17: 149 - 371°C (300 - 700°F)

FIGURE 4-9 Variation of product quality with feedstock type for a hydrocracking operation. (From G. D. Hobson and W. Pohl, eds., Modern Petroleum Technology, Applied Science Publishers, 1973, p. 314.)

4-6 DEASPHALTING

The large quantities of residua and heavy oils now being used as refinery feedstocks present problems that were minimal (if they were present at all) when lighter crude oils were the predominant refinery feedstock. For example, in catalytic cracking, the asphaltic (coke-forming) constituents of the feedstock should be preferably maintained at a minimum since the capacity of the process may actually be limited by the carbon-burning rate of the regenerating vessel. Furthermore, care should be taken to avoid catalyst poisoning or reduction in catalyst activity, and the nitrogen-containing organics (as well as metals such as vanadium and nickel) should be removed from the feedstock as nearly completely as possible. Accordingly, a number of processes have been designed for this specific purpose (described earlier in this chapter) and are directed at upgrading catalytic-cracking feedstocks in the above respects.

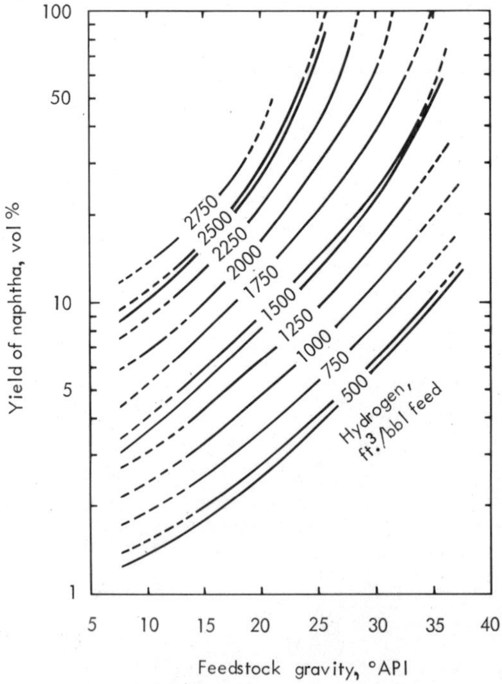

FIGURE 4-10 Hydrogen requirements for various feedstock types to produce a light naphtha. [From W. L. Nelson, Oil and Gas, 1967, 65(26): 85.]

TABLE 4-11 Feedstock Types and Product Data for a Hydrocracking Operation

Feedstock properties					
Feed Source	Kuwait	Khafji	Kuwait	Khafji	Tia Juana
Cut Range, °F	620+	700+	700/1100	700/1100	825/1050
Gravity, °API	16.0	10.9	19.9	20.9	20.1
Sulfur, wt %	3.94	4.25	3.25	3.05	1.80
Conradson carbon, wt %	9.3	14.2	–	–	–
Metals, ppm					
V	43	99	–	–	–
Ni	13	32	–	–	–
Average yields and qualities					
Light ends, wt % on FF					
$\leqslant C_4$	3.7	4.1	3.5	3.5	2.5
Naphtha (C_5-400°F)					
Yield, vol %	1.4	1.4	1.7	1.7	1.7
Gravity, °API	45.0	45.0	45.0	45.0	45.0
Sulfur, wt %	0.02	0.02	0.02	0.02	0.02
Gas oil (>400°F)					
Yield, vol %	100.1	100.4	99.6	99.5	98.6
Gravity, °API	22.2	17.5	26.2	26.9	23.1
Sulfur, wt %	1.00	1.00	0.50	0.48	0.18
H_2 Consumption, scf/bbl	600	660	420	400	325

There is, however, one other process which is capable of removing all, or a substantial fraction, of the asphaltics (asphaltenes and resins) from feedstocks that are destined for further processing. The method

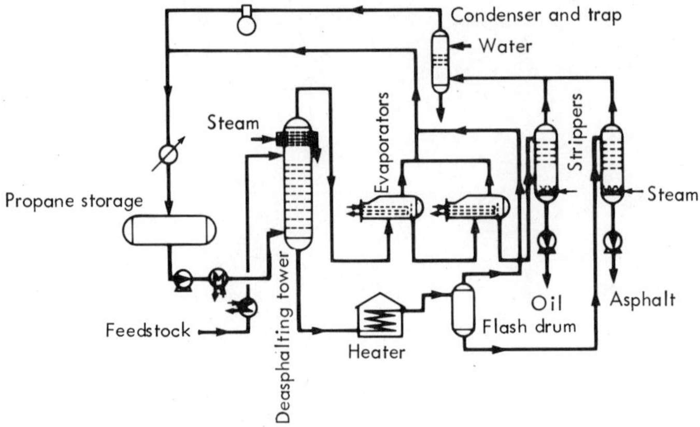

FIGURE 4-11 A propane deasphalting unit.

involves use of a liquid hydrocarbon solvent [usually the lighter hydrocarbons of the propane, butane(s) or pentane(s) group either singly or as mixtures] whereby the solvent causes separation of the asphaltic constituents from the feedstock leaving a deasphalted oil that contains substantially less sulfur and metals than the feedstock. On a commercial scale, propane is the most common solvent (Fig. 4-11) and the units are designed to operate at 40 to 80°C (105 to 175°F) and at pressures of 400 to 550 psi with the solvent/feedstock ratio usually falling in the range 5:1 to 13:1. Extraction takes place countercurrently in a tower equipped with contacting trays, and some degree of selectivity of the solvent is possible since propane, contrary to the behavior of the majority of solvents, dissolves less asphaltic material at higher temperatures.

Thus, propane deasphalting (or modifications of this process) is most efficient in improving cracking feedstocks because the asphaltic (coke-forming) constituents and the metals content are both lower in the deasphalted oil than in the original residuum or heavy oil (Table 4-12). There are claims that installation of a deasphalting unit just prior to a catalytic cracking (hydrocracking) unit may actually improve the yield and quality of the products from the cracker. There is some dispute over claims of this type, but there is no dispute that a deasphalter will improve the quality of the feedstock from the aspect of metals content and sulfur content (Table 4-13; Fig. 4-12). But, unless there are uses for the deasphalted oil as a fuel oil or as a blending stock for heavy fuel oils, there still remains the need to remove the remaining sulfur if the deasphalted oil is ultimately intended for use as one of the lighter liquid fuels.

TABLE 4-12 Feedstock and Product Stream Characteristics from a Deasphalting Operation

	Murban	Light Arabian	Kuwait	Buzurgan	Boscan
Feedstock					
Yield on crude, wt %	10	21.3	28.0	52.0	77.8
Specific gravity	0.982	1.003	1.021	1.051	1.037
API gravity	12.6	9.6	7.1	3.1	4.9
Viscosity at 210° F cSt	195	345	1105	3355	5250
Sulfur, wt %	3.03	4.05	5.35	6.20	5.90
Nickel, ppm	17	19	33	76	133
Vanadium, ppm	26	61	87	233	1264
			Solvent		
	C_3	C_3	C_3	C_3	C_3
Deasphalted oil					
Yield on residuum, wt %	54	45	29	31	33
Specific gravity	0.924	0.933	0.937	0.945	0.953
API gravity	21.6	20.2	19.5	18.2	17.0
Viscosity at 210° F cSt	38	34.9	46.0	30.0	33.0
Conradson carbon, wt %	2.0	1.65	1.90	2.0	1.8

TABLE 4-12 (Continued)

	Murban	Light Arabian	Kuwait	Buzurgan	Boscan
	Solvent				
	C_3	C_3	C_3	C_3	C_3
Deasphalted oil (continued)					
Asphaltenes, wt %	<0.05	<0.05	<0.05	<0.05	<0.05
Nickel, ppm	<1	1.0	1.0	2.0	6.0
Vanadium, ppm	<1	1.4	2.0	4.0	10.0
Sulfur, wt %	1.80	2.55	2.70	3.60	4.70
Nitrogen wt %	0.08	0.12	0.13	0.09	0.16
	Solvent				
	C_5	C_5	C_5	C_5	C_5
Deasphalted oil					
Yield on residuum, wt %	93.0	85.5	78	64.2	64.0
Specific gravity	0.970	0.974	0.986	0.985	0.987
API gravity	14.4	13.8	12.0	12.2	11.9
Viscosity at 210° F cSt	124	105	201	78	135

Conradson carbon, wt %	9	7.9	8.9	7.4	6.4
Asphaltenes, wt %	<0.05	<0.05	<0.05	<0.05	<0.05
Nickel, ppm	8	7	12	12	36
Vanadium, ppm	9	15.5	28	26	270
Sulfur, wt %	2.85	3.65	4.70	4.80	5.30
Nitrogen, wt %	0.16	0.22	0.21	0.24	0.43

Source: A. Billon, J. P. Peries, E. Fehr, and E. Lorenz, Oil and Gas, Jan. 24, 1977, p. 45.

TABLE 4-13 Feed Component Left in Deasphalted Oils from Several Residua

	Sulfur			Nitrogen			Nickel			Vanadium		
Deasphalted oil yield, vol %	70	80	90	70	80	90	70	80	90	70	80	90
Component left in oil, wt %												
Gach Saran	50	68	83	33	48	70	17	32	57	11	26	55
Kuwait	57	68	83	38	54	74	6	15	38	7	19	44
Iranian Light	53	66	81	34	51	71	10	19	41	6	12	28
Tia Juana	50	64	80	34	50	71	5.5	14.5	38	3.5	10	33

Source: J. G. Ditman, Hydrocarbon Processing, 1973, 52(5):111.

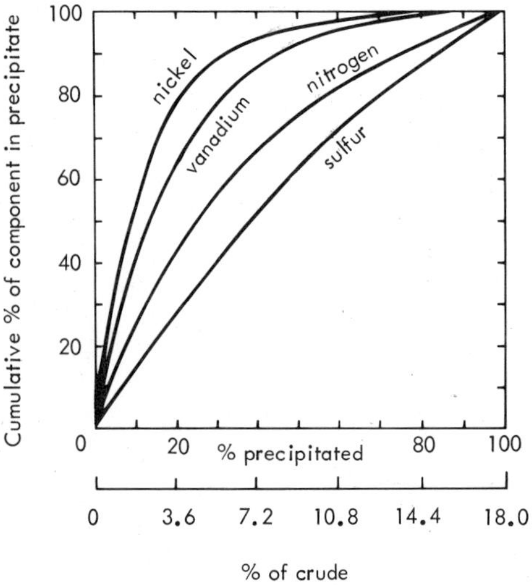

FIGURE 4-12 Distribution of the heteroatoms and metals in the precipitate after successive treatments of a Kuwait residuum with propane, butane, pentane, and heptane. [From R. A. Flynn, H. Beuther, and B. K. Schmid, Petroleum Refiner, 1961, 40(4):141.]

BIBLIOGRAPHY

Billon, A., Franck, J. P., Peries, J. P., Gallei, E., Fehr, E., and Lorenz, E.: More ways to use hydrocracking, Hydrocarbon Processing, 122, May 1978.

Billon, A., Lorenz, E., and Peries, J. P.: Valorisation of crude bottoms, Proc. Third International Symposium on Large Chemical Plants, New Technologies, and Varying Feedstocks, Antwerp, 115, October 20-21, 1976.

Billon, A., Peries, J. P., Fehr, E., and Lorenz, E.: SDA key to upgrading heavy crudes, Oil and Gas Journal, 43, January 24, 1977.

Bland, William F., and Davidson, Robert F. (eds.): Petroleum Processing Handbook, McGraw-Hill, New York, 1967.

Blaser, D. E., Rionda, J. A., and Saxton, A. L.: Combine desulphurising and coking, Hydrocarbon Processing, 137, September 1971.

Ditman, John G.: Deasphalt to get feed for lubes, Hydrocarbon Processing, 110, May 1973.

Edison, R. R., Siemssen, J. O., and Masologites, G. P.: Crude and resid can be cat-cracker feeds, Oil and Gas Journal, 54, December 20, 1976.

Flinn, R. A., Beuther, H., and Schmid, B. K.: Now you can improve residue treating, Petroleum Refiner, 40(4):139, 1961.

Gary, James H., and Handwerk, Glenn E.: Petroleum Refining: Technology and Economics, Marcel Dekker, New York, 1975.

Gearhart, J. A., and Garwin, L.: ROSE process improves resid feeds, Hydrocarbon Processing, 125, May 1976.

Gearhart, J. A., and Garwin, L.: Resid-extraction process offers flexibility, Oil and Gas Journal, 63, June 14, 1976.

Gould, Robert F. (ed.): Origin and Refining of Petroleum, Advances in Chemistry Series No. 103, American Chemical Society, Washington, D.C., 1971.

Hobson, G. D., and Pohl, W (eds.): Modern Petroleum Technology, Applied Science Publishers, Barking, Essex, England, 1973.

McKinney, C. M.: Sulphur in products vs. crude oil, Hydrocarbon Processing, 117, October 1972.

Martin, S. W., and Willis, L. E.: Coking petroleum residues, in Advances in Petroleum Refining, Volume 2, 1959, Chapter 6.

Miyake, T., Mori, T., Horii, Y., and Kanda, T.: Cherry P process—treatment of heavy fractions, Proc. Pacific Chem. Eng. Cong., 2(1):557, 1977.

Murphy, J. R., and Soudek, M.: Modern trends in fluid catalytic cracking, Proc. Third International Symposium on Large Chemical Plants, New Technologies, and Varying Feedstocks, Antwerp, 81, October 20-21 1976.

Nelson, Wilbur L.: Petroleum Refinery Engineering, McGraw-Hill, New York, 1958.

Nelson, Wilbur L.: Sulphur reduction by visbreaking, Oil and Gas Journal, 101, August 21, 1967.

Nelson, Wilbur L.: Can sulphur be reduced by visbreaking?, Oil and Gas Journal, November 6, 1967.

Nelson, Wilbur L.: Low-sulphur fuel oils—2, Oil and Gas Journal, 76, November 26, 1973.

Ondish, G. F., Frayer, J. A., and McKinney, J. D.: Save crude—feed resin to FCC, Hydrocarbon Processing, 105, July 1976.

Rhoe, A., and de Blignieres, C.: Visbreaking: A flexible process, Hydrocarbon Processing, 131, January 1979.

Rossi, W. J., Deighton, B. S., and MacDonald, A. J.: Get more light fuel from resid, Hydrocarbon Processing, 105, May 1977.

Sikonia, John G., Jacobs, William L., and Gembicki, Stanley A.: Hydrocrack for more distillates, Hydrocarbon Processing, 117, May 1978.

Sulphur distribution in extreme or freak crude oils, Oil and Gas Journal, 92, December 4, 1967.

Viloria, D. A., Krasuk, J. H., Rodriguez, O., Buenatama, H., and Lubkowitz, J.: Solvent deasphalt boscan crude, Hydrocarbon Processing, 109, March 1977.

Ward, John W.: The varieties of hydrocracking, Hydrocarbon Processing, 101, September 1975.

Washimi, K., Kaji, H., Ozaki, K., and Nakanishi, H.: New thermal cracking for vacuum residue (Eureka process), Proc. Pacific Chem. Eng. Cong., 2(1):534, 1977.

Wollaston, E. G., Haflin, W. J., Ford, W. D., and D'Sonza, G. J.: What influences cat cracking?, Hydrocarbon Processing, 93, September 1975.

5

THE HYDRODESULFURIZATION PROCESS

5-1 PROCESS DEFINITION

The hydrodesulfurization process is actually a specific hydrogenation process and one of several that might be applied to the conversion of petroleum fractions or to the purification of petroleum products. Hydrogenation processes, as employed in petroleum refining, can be classified as nondestructive or destructive. Nondestructive, or simple, hydrogenation is generally used for the purpose of improving product quality without appreciable alteration of the boiling range. Mild processing conditions are employed so that only the more unstable materials are attacked and the sulfur, nitrogen, and oxygen compounds undergo hydrogenolysis to split out hydrogen sulfide, ammonia, and water, respectively.

On the other hand, destructive hydrogenation (hydrocracking) is characterized by the cleavage of carbon-carbon bonds with concurrent addition of hydrogen to the fragments to produce saturated lower-boiling products. Such treatment requires severe reaction temperatures (usually of the order of 360 to 410°C, 680 to 770°F) as well as high hydrogen pressures to minimize reactions which lead to the formation of undesirable products such as coke.

Thus, the hydrodesulfurization of various feedstocks can fall into the category of nondestructive hydrogenation or destructive hydrogen. Examples of the former case would be the hydrodesulfurization of naphthas with temperatures of the order of 330 to 370°C (625 to 695°F) and hydrogen partial pressures of 100 to 500 psi. In contrast, and more pertinent to the present text, destructive hydrogenation could well be employed as a means of the hydrodesulfurization of the heavier feedstocks such as the residua and heavy oils. In this case, desulfurization would still be a major object of the process but, an additional factor, the conversion of the feedstock to lower-boiling products may also be required. In such a process, the tendency to coke formation (from the higher-molecular-weight constituents of the feedstock) is very strong and minimizing this tendency can be achieved by using higher pressures (of the order of 750 to 2000 psi) than used for a naphtha, while higher temperatures (360 to 410° C, 680 to 770° F) will most certainly be required to convert the high-molecular-weight components of the feedstock to lower-boiling hydrogenated products through the agency of carbon-carbon bond scission. In essence, the feedstock plays an important part in determining the hydrodesulfurization conditions (Chapter 6).

Thus, on a chemical basis, it is possible to define the hydrodesulfurization process in terms of the feedstock type and the required products. In practice, it may actually be difficult, if not impossible, to carry out nondestructive hydrogenation and completely eliminate carbon-carbon bond scission as in the case of the desulfurization of naphthas while for residua and heavy oils it is usually preferable to promote hydrocracking as an integral part of the desulfurization process.

5-2 PROCESS DESCRIPTION

The hydrodesulfurization process is essentially the reaction of hydrogen with a predominantly hydrocarbon feedstock to produce a desulfurized hydrocarbon product and hydrogen sulfide (Chapter 3). In a very simplified process (Fig. 5-1), the feedstock is first pressurized to a pressure which is a little higher than that of the reactor section, mixed with hot recycle gas and preheated to the temperature of the reactor inlet. The hot feedstock (and the recycle gas) is then introduced to the catalyst in the reactor where temperatures of the order of 290 to 455°C (550 to 850°F) and pressures in the range 150 to 3000 psi prevail. Heat exchangers are employed to cool the reactor effluent and the desulfurized liquid product is separated from the recycle gas at a pressure somewhat lower than that of the reactor

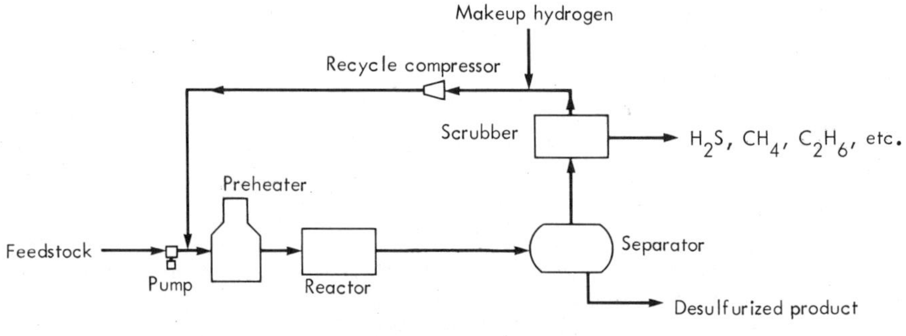

FIGURE 5-1 Schematic representation of the hydrodesulfurization process.

section. Hydrogen sulfide and any light hydrocarbon gases are removed from the recycle gas which is then mixed with fresh (makeup) hydrogen, recompressed, and mixed with further hydrocarbon feedstock.

The use of a recycle gas technique in the hydrodesulfurization process minimizes losses of hydrogen—a very expensive commodity in petroleum refining. The hydrodesulfurization process requires high partial pressures of hydrogen in order to promote high desulfurization reaction rates and to diminish coke (carbon) deposition on the catalyst. In order to maintain high partial pressures of hydrogen, it is necessary to introduce hydrogen into the reactor at several times the rate of hydrogen consumption, and it is possible to recover most of the unconsumed hydrogen in the separator after which it is then recycled for further use. Indeed, if it were not possible to recover most of the unconsumed hydrogen, the process economics would probably be very questionable. However, tolerable amounts of hydrogen may be lost as a result of the solubility of the gas in the liquid hydrocarbon product or even during the removal of the hydrogen sulfide (and the light hydrocarbon gases) from the recycle gas.

Hydrogen for the hydrodesulfurization process has to be produced on-site and would usually come from a hydrogen plant which may utilize the concept of steam-methane reforming (see Chapter 8) or a similar process involving the generation of hydrogen from a low-molecular-weight hydrocarbon. Alternatively, the hydrogen could be produced as a by-product from one of the many catalytic reforming processes that are available to refinery operators. These are the processes by which high-octane gasolines are produced from low-octane naphthas. The reactions involved in this type of process are extremely complex, but the overall effect of a reforming process is that the change in the boiling point of the feedstock passed through a reforming unit is relatively small as it is not the intent of the process to induce severe thermal degradation of the components of the charge stock but to rearrange the structure of the components and so produce higher-octane gasolines. During the reforming process, dehydrogenation of certain molecular types occurs with the production of hydrogen gas which is then available for any one or more of the processes that require extraneous hydrogen. However, depending upon the needs of any particular refinery, it may be necessary to have a hydrogen production plant to produce hydrogen additional to that which may be produced by a reforming unit.

Hydrogen requirements for the hydrodesulfurization process (Table 5-1) depend upon the nature of the feedstock as well as on the extent of the desulfurization. For example, the heavier feedstocks require substantially more hydrogen to produce a given product than the lighter feedstocks (Fig. 4-10) or to produce a product with a predetermined amount of sulfur (Figs. 5-2 and 5-3). However, it should be noted that the theoretical hydrogen requirements often differ markedly from the experimental values (Fig. 5-4) because of the nitrogen and oxygen contents of the feedstock (as these atoms are removed during the process as their respective hydrogen analogs—

TABLE 5-1 Hydrogen Consumption (Chemical) During the Processing of Various Residua and Heavy Oils

	°API	Sulfur (wt %)	Carbon residue (Conradson) (wt %)	Nitrogen (wt %)	Hydrogen (scf/bbl)	Desulfurization (%)
Venezuela, atmospheric	15.3–17.2	2.1–2.2	9.9–10.4	–	425–730	48–86
Venezuela, vacuum	4.5–7.5	2.9–3.2	20.5–21.4	–	825–950	47–60
Boscan (whole crude)	10.4	5.6	–	0.52	1100	70
Tia Juana, vacuum	7.8	2.5	21.4[a]	0.52	490–770	39–67
Bachaquero, vacuum	5.8	3.7	23.1[a]	0.56	1080–1260	59–73
West Texas, atmospheric	17.7–17.9	2.2–2.5	8.4	–	520–670	80–86
West Texas, vacuum	10.0–13.8	2.3–3.2	12.2–14.8	–	675–1200	55–92
Khafji, atmospheric	15.1–15.7	4.0–4.1	11.0–12.2	–	725–800	71–88
Khafji, vacuum	5.0	5.4	21.0	–	1000–1100	68–78
Arabian light, vacuum	8.5	3.8	–	–	435–1180	53–92
Kuwait, atmospheric	15.7–17.2	3.7–4.0	8.6–9.5	0.20–0.23	470–815	63–92
Kuwait, vacuum	5.5–8.0	5.1–5.5	16.0	–	290–1200	30–95

[a] Ramsbottom carbon residue.

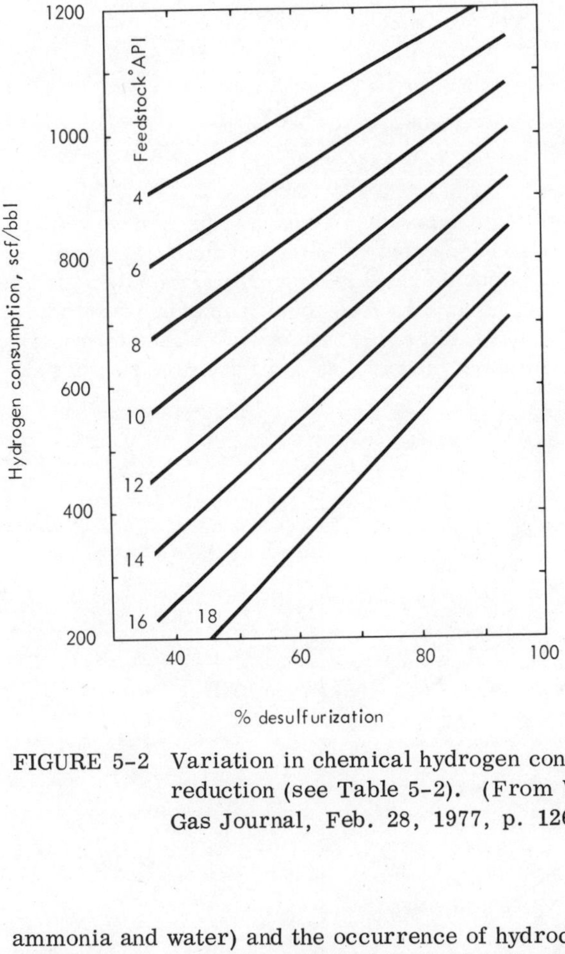

FIGURE 5-2 Variation in chemical hydrogen consumption with sulfur reduction (see Table 5-2). (From W. L. Nelson, Oil and Gas Journal, Feb. 28, 1977, p. 126.)

ammonia and water) and the occurrence of hydrocracking during the desulfurization. Metals content of the feedstock may also influence the hydrogen consumption by altering the characteristics of the catalyst—it is conceivable that metals deposition onto the catalyst may serve to increase the hydrocracking activity of the catalyst to the detriment of the hydrotreating activity, thereby promoting higher consumption of hydrogen in the process (Table 5-2).

Sulfur removal by a hydrodesulfurization process is usually good to excellent and may even be of the order of 90 to 95%. The products formed by the removal of sulfur from the various molecular types that may make up any of a variety of feedstocks usually are considerably more volatile than the parent sulfur compounds:

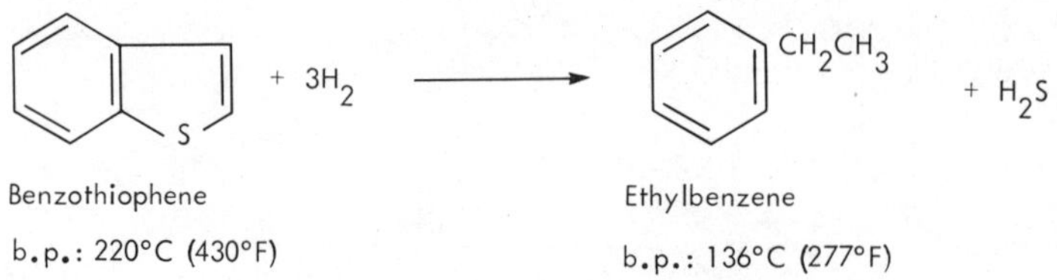

and certain of these low-boiling compounds may have to be removed from the product mix to maintain an acceptable volatility. This is especially true where the feedstock to be desulfurized is a naphtha or middle distillate (light fuel oil); in the case of the naphtha feedstock, it may be preferable that the gasoline product have a small proportion of low-boiling hydrocarbons present to promote ignition of the gasoline while the presence of these

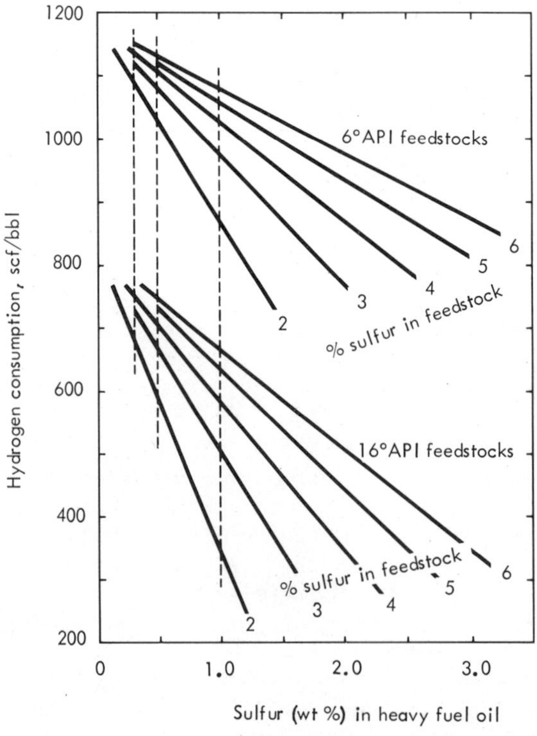

FIGURE 5-3 Variation in chemical hydrogen consumption to produce a heavy fuel oil of given sulfur content (see Table 5-2). (From W. L. Nelson, Oil and Gas Journal, Feb. 28, 1977, 1. 127.)

TABLE 5-2 Additions to Chemical Hydrogen Consumption Data (Figs. 5-2, 5-3) for Metals and Nitrogen Compounds in the Feedstock

Metals	
V + Ni (ppm)	Corrections (%) to hydrogen consumption
0-100	-2
200	1
300	2.5
400	4
500	6.5
600	9
700	12
800	16
900	21
1000	28
1100	38
1200	50

Nitrogen Compounds	Additional hydrogen required H_2/compound (mol)	scf/bbl Feed
Saturated amines	1	83
Pyrrolidine	2	167
Nitriles, pyrroline, alkyl cyanides	3	250
Pyrrole, nitroparaffins	4	334
Analine, pyridine	5	417
Indole	7	584

Source: W. L. Nelson, Oil and Gas Journal, Feb. 28, 1977, p. 127.

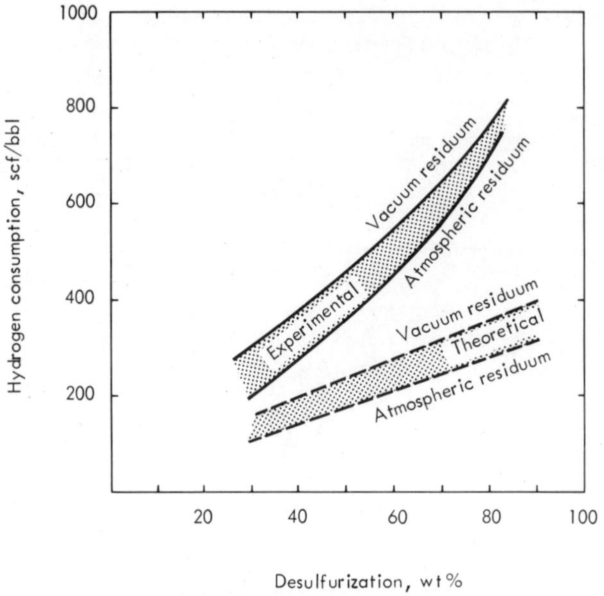

FIGURE 5-4 Chemical hydrogen consumption for the hydrodesulfurization of Kuwait residua.

same hydrocarbons in a fuel oil might increase the flash point to such an extent that the fuel oil be classified as unstable and unfit for the specified use. On the other hand, desulfurization may increase the aromatics content of the product stream relative to the feedstock—a very desirable property where an increase in the octane rating is sought. In short, the desulfurization process may need careful monitoring when feedstock type can have a marked influence on the properties of the product and downstream treatments may be necessary to ensure that product specifications are met.

In the same way, the desulfurization of heavy oils and residua may also require careful monitoring of the product properties. However, the nature of the feedstock may preclude the production of high-volatile materials if the sole purpose of the process is to reduce the sulfur content of the feedstock, thereby producing the equivalent of a heavy gas oil feedstock. If, however, the process conditions are such that hydrocracking becomes one of the major chemical reaction types, a high proportion of low-boiling material may be found in the product mix and have to be removed accordingly prior to any further processing of the product stream.

Perhaps the major advantage of the hydrodesulfurization process is the fact that the sulfur is ultimately removed as hydrogen sulfide which may, by virtue of its gaseous nature, be readily and completely removed from the hydrocarbon product streams. The manner in which the hydrogen sulfide is removed (as a process waste or as a precursor for elemental sulfur manufacture) is in direct contrast to the problems that may arise when other methods of desulfurization are employed. For example, desulfurization of a feedstock by any of the thermal methods where extraneous hydrogen is not employed (Chapter 4) merely tends to concentrate the sulfur in the coke (or, if the process is distillation, in the residuum) that is formed, thereby merely postponing, rather than dispelling, the problem. Indeed, the sulfur still remains in one of the process products and will eventually have to be removed if the product is to satisfy specifications. Furthermore, it is doubtful if any one of these alternate methods will ever reach a satisfactory level of desulfurization compared to the hydrodesulfurization process where desulfurization levels of at least 80% are desirable, and, in the majority of cases, are often recorded.

The hydrodesulfurization of high-boiling feedstocks (such as residua and heavy oils) which have sulfur contents up to about 5% by weight are similar to those employed for gas oils but with residua, and heavy oil feedstocks, the tendency for coke (carbon) deposition on the catalyst and contamination of the catalyst by the metals in the feedstock is much higher. Consequently, unless a low level of desulfurization is acceptable, it is necessary to regenerate the catalyst at frequent intervals. Furthermore, these heavier feedstocks contain appreciable amounts of asphaltics and the sulfur compounds are a significant proportion of the total compounds present.

5-3 REACTOR DESIGN

Reactor designs for hydrodesulfurization of various feedstocks vary in the way in which the feedstock is introduced into the reactor and in the arrangement, as well as the physical nature, of the catalyst bed. The conditions under which the hydrodesulfurization process operates (i.e., high temperatures and high pressures) dictate required wall thickness (determined by the pressure/temperature/strength ratio). In addition, resistance of the reactor walls to the corrosive attack by hydrogen sulfide and hydrogen (to name only two of the potential corrosive agents of all of the constituents in, or arising from, the feedstock) can be a problem and precautions should be taken to ensure that wall thickness and composition yield maximum use and safety.

With these criteria in mind, various reactors have been designed to satisfy the needs of the hydrodesulfurization process. Thus, reactors may vary from as little as 4 ft in diameter to as much as 20 ft in diameter and have a wall thickness anywhere from 4.5 to 10 in. or so. These vessels may weigh from 150 tons to as much as 1000 tons. Obviously, before selecting a suitable reactor, shipping and handling requirements (in addition to the more conventional process economics) must be given serious consideration.

It is beyond the scope of the present text to consider mechanical design and material selection, but it is the purpose of this section to describe the types of reactor available and the means by which the feedstock and hydrogen are introduced to the catalyst.

A. The Downflow Fixed-Bed Reactor

The downflow fixed-bed reactor has been used widely for hydrodesulfurization processes and is so called because of the feedstock entry at the top of the reactor while the product stream is discharged from the base of the reactor (Fig. 5-5). The catalyst is contained in the reactor as stationary beds with the feedstock and hydrogen passing through the bed in a downward direction. The exothermic nature of the reaction and the subsequent marked temperature rise from the inlet to the outlet of each catalyst bed require that the reaction mix be quenched by cold recycle gas at various points in the reactor and hence the incorporation of separate catalyst beds as part of the reactor design.

To combat the inevitable loss in desulfurizing activity of the catalyst which must be presumed to occur with time under any predetermined set of reaction conditions, the bed inlet temperature may be increased slowly, thereby increasing the overall temperature of the catalyst bed and so maintaining constant catalyst activity. Thus, depending upon the nature of the feedstock, there may be considerable difference between the start-of-run temperature and the end-of-run temperature.

In fixed-bed reactors, the catalyst may be poisoned (deactivated) progressively. For example, the first catalyst bed will most likely be poisoned by vanadium and nickel deposition initially at the inlet to the bed and then progressively through the bed as the active zone is gradually pushed downward into the bed. Once catalyst poisoning has progressed through the bed, the catalyst may have to be discarded or regenerated. However, catalyst regeneration may only suffice for a limited time and the catalyst may have to be completely replaced. This is especially true in the case of feedstocks such as residua and heavy oils—removal of the metal contaminants is not usually economical, or efficient, during rapid regeneration. In fact,

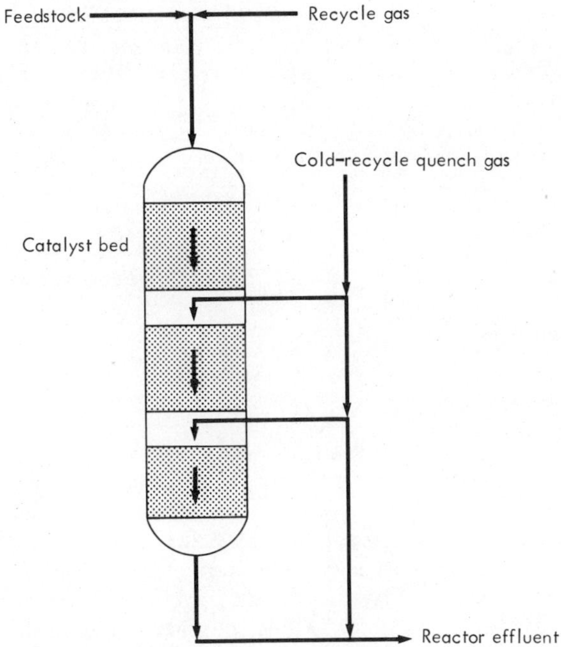

FIGURE 5-5 A fixed-bed downflow hydrodesulfurization reactor.

the deposited metals are believed to form sulfates during removal of carbon and sulfur compounds by combustion which produce a permanent poisoning effect. Thus, if fixed-bed reactors are to be used for residuum or heavy oil hydrodesulfurization (in place of the more usual distillate hydrodesulfurization) it may be necessary to first process the heavier feedstocks to remove the metals (especially vanadium and nickel) and so decrease the extent of catalyst bed plugging. Precautions should also be taken to ensure that plugging of the bed does not lead to the formation of channels within the catalyst bed which will also reduce the efficiency of the process and may even lead to pressure variances within the reactor because of the distorted flow patterns with eventual damage.

B. The Radial-flow Fixed-bed Reactor

The radial-flow fixed-bed hydrodesulfurization reactor is a variance of the downflow fixed-bed reactor. Again, the feedstock enters the top of the

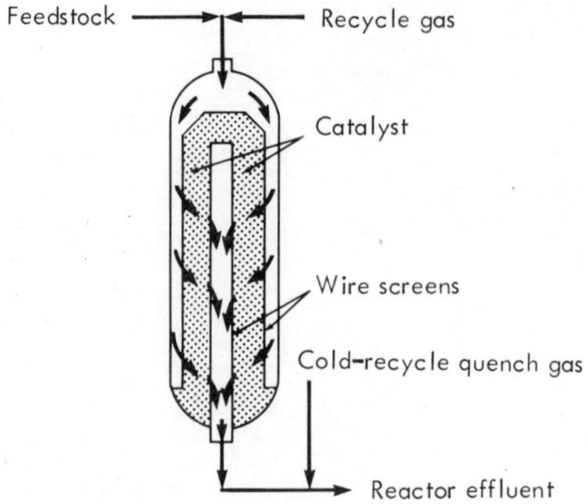

FIGURE 5-6 A fixed-bed radial-flow hydrodesulfurization reactor.

reactor but instead of flowing downward through the catalyst bed, the feedstock is encouraged to flow through the bed in a radial direction (Fig. 5-6) and then out through the base of the reactor. There are certain advantages of this type of reactor, not the least of which is a low pressure drop through the catalyst bed; in addition, a radial bed reactor has a larger catalyst cross-sectional area as well as a shorter bed depth than the

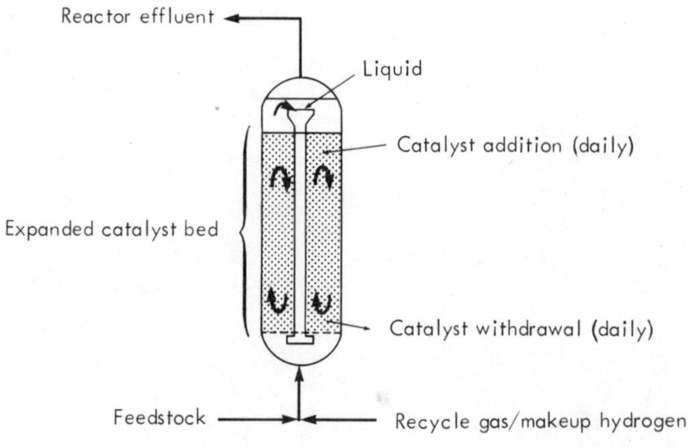

FIGURE 5-7 An expanded (particulate fludized) bed hydrodesulfurization reactor.

corresponding downflow reactor. It is this latter property which gives rise to the smaller pressure drop across the catalyst bed. However, there are more chances of localized heating in the catalyst bed and (in addition to the more expensive reactor design per unit volume of catalyst bed) it may be more difficult to remove contaminants from the bed as part of the catalyst regeneration sequence. For this reason alone, it is preferable that this type of reactor be limited to hydrodesulfurization of low-boiling feedstocks such as naphthas and kerosenes and application to the higher-boiling heavy oils and residua is usually not recommended.

C. The Upflow Expanded-bed Reactor (Particulate Fluidized-bed Reactor)

Expanded-bed reactors operate in such a way that the catalyst remains loosely packed and is less susceptible to plugging and they are therefore more suitable for the heavier feedstocks as well as for feedstocks that may contain considerable amounts of suspended solid material. Because of the nature of the catalyst bed, such suspended material will pass through the bed without causing frequent plugging problems. Furthermore, the expanded state of motion of the catalyst allows frequent withdrawal from, or addition to, the catalyst bed during operation of the reactor without the necessity of shutdown of the unit for catalyst replacement. This property alone makes the ebullated reactor ideally suited for the high-metal feedstocks (i.e., residua and heavy oils) which rapidly poison a catalyst with the ever-present catalyst replacement problem (Fig. 5-7).

In the reactor, the feedstock and hydrogen are caused to flow upward through the catalyst bed in which each catalyst particle is reputed to have independent motion and can therefore (in theory) migrate throughout the entire catalyst bed. The heat of the reaction (which could be a problem in other reactors if it were not controlled) is dissipated in raising the temperature of fresh feedstock to that existing in the reactor. Spent catalyst can be withdrawn from the reactor and replaced with fresh catalyst on a daily basis if necessary and the need for time-wasting and costly shutdowns can thus be eliminated.

Expanded-bed reactors normally operate with 1/32-in. extruded catalysts which require partial recycle of the liquid products to maintain the catalyst particles in an expanded and fluidized state.

The near completely random motion of the catalyst bed virtually ensures an isothermal operation, but the efficiency of the hydrodesulfurization reaction tends to suffer because of the back-mixing of the product and feedstock. Hence, to effect sulfur removal at over 75% efficiency, it may be necessary to operate with two or more reactors in series. The need for two or more of these units to effectively desulfurize a feedstock may be cited as a disadvantage of the reactor, but the ability of the reactor to operate under isothermal conditions as well as the onstream catalyst addition-withdrawal system and the fact that the reactor size required for an expanded catalyst bed is often smaller than that required for a fixed bed can be cited in support of such a unit.

It is also possible to use finely divided catalyst in the expanded-bed reactor. If the catalyst is a suitable size (50 to 200 μm, i.e., 50×10^{-4} to 200×10^{-4} cm), it is possible to operate the expanded-bed reactor without recycling the liquid products to maintain the catalyst in the fluidized state. In addition, the finely divided catalyst has a relatively larger number of external pores on the surface than the extruded catalyst and is less likely to have metal contaminants plugging up these pores because of their size. The overall effect of a finely divided catalyst in this reactor is more efficient sulfur removal for a given set of process conditions.

5-4 HYDRODESULFURIZATION CATALYSTS

Many of the catalysts used for the hydrodesulfurization process are formed by compositing various transition metals (or their salts) with a solid support, with the metal constituent being the "active" catalyst. The most commonly used materials for supports are alumina, silica, silica-alumina, kieselguhr, magnesia (and other metal oxides), as well as the zeolites. The support can be manufactured in a variety of shapes or may even be crushed to particles of the desired size. The metal constituent can then be added by contact of the support with an aqueous solution of the metal salt and the whole subjected to further treatment which will dictate the final form of the metal on the support (i.e., the metal oxide, the metal sulfide, or even the metal itself).

The surface area of the catalyst is usually large (200 to 300 $m^2\ g^{-1}$), but almost all of this surface is contained within the pore space of the alumina. Cobalt and molybdenum are two of the more common metals that are used as hydrodesulfurization catalysts and, as such, are dispersed in a thin layer within the pore system of the alumina. When these metals are used together as a hydrodesulfurization catalyst, the catalyst is more tolerant to poisoning agents and is usually classed as being suitable for a wide variety of feedstocks but more particularly to the heavy oils and residua. Other metals may be used such as a combination of nickel and molybdenum, but this catalyst is a more active hydrogenation catalyst than the cobalt-molybdenum catalyst and consumes more hydrogen per mole of sulfur removed. However, the nickel-molybdenum catalyst is useful for the hydrodesulfurization of catalytic cracking feedstocks (where maximum

TABLE 5-3 Reaction Types and Relative Reactivities of the Common Hydrodesulfurization Catalysts

Catalyst[a]	Hydrodesulfurization	Hydrodenitrogenation	Aromatics hydrogenation
Co–Mo–Al_2O_3	xxxx[b]	xx	x
Ni–Mo–Al_2O_3	xxx	xxx	xx
Ni–W–Al_2O_3	xx	xx	xxxx

[a] All catalysts are in the "sulfide" form with the average chemical composition CoO (or NiO) 2-4%; MoO_3 12-15%; and Al_2O_3 81-86%.
[b] Reactivities: x–fair; xx–good; xxx–very good; xxxx–excellent.

hydrogen consumption is desirable) and it is more selective for nitrogen removal from feedstocks and presumably can tolerate higher nitrogen feedstocks without losing as much activity in a given time than the cobalt-molybdenum catalyst (Table 5-3).

Another catalyst that has received some attention is nickel-tungsten on alumina, which is a very active hydrogenation catalyst and displays a high activity in hydrocracking reactions.

The hydrodesulfurization catalysts are normally used as extrudates or as porous pellets, but the particle size and pore geometry have an important influence on process design—especially for the heavier feedstocks (see above)—since the reaction rates of hydrodesulfurization catalysts are limited by the diffusion of the reactants into, and the products out of, the catalyst pore systems. Thus, as the catalyst particle size is decreased, the rate of desulfurization is increased (Fig. 5-8) but the pressure differential across the catalyst bed also diminishes and a balance must be reached between reaction rate and pressure drop across the bed.

Hydrodesulfurization catalysts are usually more active in the sulfided form with the external sulfur being applied either during the preparation from a sulfur-containing source or during the initial contact of the catalyst with the sulfur-containing feedstock. Chemically, the sulfiding process is a reduction in the oxide form of the metal (cobalt, nickel, etc.) on the alumina support by conversion to the metal sulfide. This can be achieved, in the absence of a sulfur-containing feedstock, by injecting hydrogen sulfide or any other low-boiling sulfide (carbon disulfide, dimethyl sulfide, and the like) directly into the recirculating hydrogen stream. If the

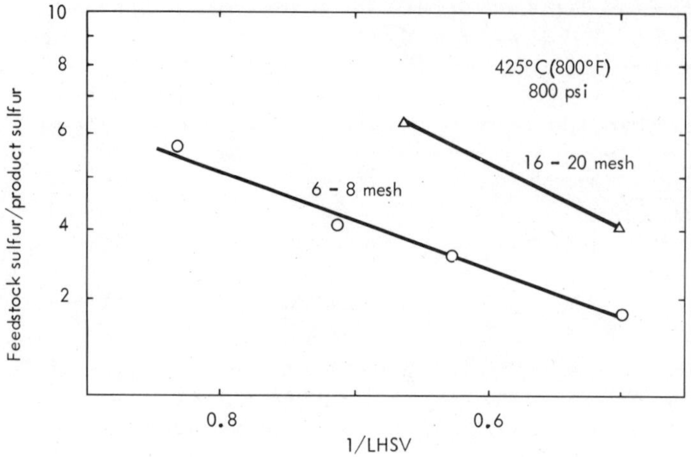

FIGURE 5-8 Effect of catalyst particle size on the desulfurization of a topped Venezuelan crude oil. (From C. M. Frost and P. L. Cottingham, Bureau of Mines Report of Investigations, 1971, R1 7557, p. 8.)

feedstock is present and contains above 1% sulfur, addition of an external source is not necessary. Temperatures for the sulfiding process are usually of the order of 290 to 315°C (550 to 600°F), while minimum pressures of about 150 psi are also recommended. Caution is advised against using temperatures in excess of 315°C (600°F) since the hot hydrogen gas can reduce a substantial portion of the metal oxides in the catalyst and, with respect to the sulfiding reaction, the reactivity of the reduced metals is lower than the reactivity of the metal oxides. The sulfiding process is continued until the sulfur content of the catalyst has reached a predetermined level—usually estimated by the level of hydrogen sulfide in the tail gas.

Catalyst consumption is a major aspect of the hydrodesulfurization process and costs of the process increase markedly with the high-metal feedstocks. The ease with which the catalyst can be replaced depends, to a large extent, on the bed type (see above), and with the high-metal feedstocks it is inevitable that frequent catalyst replacement will occur. From the data available (Table 5-4), attempts have been made to produce correlations (Fig. 5-9) which will indicate catalyst efficiency for the various groups of metal-containing feedstocks. However, as with all estimations of this type, they should be used only to present an indication of the trend in the hydrodesulfurization rather than as absolute data.

5-5 CATALYST BED PLUGGING

Catalyst bed plugging can arise in a variety of ways, but the overall effect of bed plugging is always the same: expensive shutdowns and possibly complete renewal of the expensive catalyst. Thus, the deposition of rust,* coke, or metal salts (e.g., sodium chloride) from heavier and dirtier

*Rust may arise from the inclusion of iron salts in the feedstock during transportation of the crude oil where loose pieces of corrosion debris may be dislodged from the pipeline or storage tank.

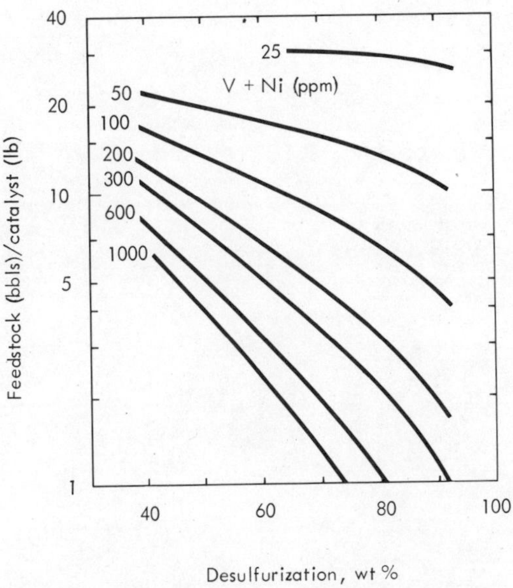

FIGURE 5-9 Estimation of catalyst consumption for various metal-containing feedstocks. (From W. L. Nelson, Oil and Gas, Nov. 15, 1976, p. 72.)

TABLE 5-4 Catalyst Consumption in the Desulfurization of Various Feedstocks

Feed	Process	°API	V + Ni (ppm)	Sulfur (wt %)	Desulfurization (%)	Catalyst consumption (bbl/lb)
Kuwait	UOP	15.8	64	4.0	75	10
Kuwait	RCD-UOP		60	3.9	88	13.7
Kuwait	Gulf HDS	16.6	60	3.8	65	17.3
Kuwait	Gulf HDS	16.6	60	3.8	88	9.3
Kuwait	RCD-UOP	16.0	62	3.9	74.5	14.5
Kuwait	RCD-UOP	16.0	62	3.9	92	11.1
Kuwait	Gulf HDS	16.6	60	3.8	87	11.9
Kuwait	Gulf HDS	16.6	60	3.8	97	6.1
Safaniya	Exxon	11.0	130	4.3	75	6.7
Arabian, light	Exxon	16.1	40	3.0	67	10.8
Kuwait, atmospheric	H-Oil	12.5	60	4.4	76	8.5
Kuwait, atmospheric	H-Oil	16.1	60	3.8	74	16.7
Kuwait, vacuum	H-Oil	7.0	120	5.5	57.5	6.1
Kuwait, vacuum	H-Oil	10.2	90	5.0	58	10
Venezuela, atmospheric	H-Oil	15.3	320	2.2	54.5	8.3
Venezuela, vacuum	H-Oil	4.5	690	2.9	52	5

Shuiba	H-Oil	10.2	90	5.0	48	11
Boscan	H-Oil	10.4	1185	5.6	70	1.8
Boscan	H-Oil	10.4	—	5.6	70	10
Boscan	H-Oil	10.4	—	5.6	84.5	6.1
Arabian, heavy	H-Oil	11.0	140	4.6	56	7.3
Tia Juana, vacuum	H-Oil	7.8	480	2.5	39	7.9
Tia Juana, vacuum	H-Oil	7.8	480	2.5	59	5.6
Bachaquero, vacuum	H-Oil	5.8	970	3.7	67	7.3
Bachaquero, vacuum	H-Oil	5.8	970	3.7	73	5.6
Gach Saran, vacuum	H-Oil	9.9	329	3.2	74	3.1
Gach Saran, vacuum	H-Oil	9.2	359	3.3	62	3.1
Gach Saran, vacuum	H-Oil	10.0	325	3.1	23	1.2
Gach Saran, vacuum	H-Oil	10.0	325	3.1	8	2.9
Salamanca	H-Oil	8.4	162	3.9	63	6.8
Salamanca	H-Oil	10.7	182	3.2	68	8.2
Salamanca	H-Oil	6.4	317	3.4	63	8.6

Source: W. L. Nelson, Oil and Gas Journal, Nov. 15, 1976, p. 73.

feedstock may all contribute to the plugging of a catalyst bed. Vanadium and nickel may also be deposited onto the surface of the catalyst as well as into the pore system. Asphaltene deposition from residua and heavy oils is also a potential means of bed plugging—coagulation of the asphaltenes becomes appreciable at temperatures above 420°C (790°F) with the formation of hard, cokelike materials on the catalyst. The exothermal nature of the reaction may also contribute indirectly to catalyst plugging—for example, lack of proper control over the heat liberated during the hydrodesulfurization process may lead to the formation of localized hot spots in the catalyst which can then initiate asphaltene coagulation at these points. In a similar manner, adsorption of the lighter constituents from a whole crude (heavy oil) or from, say, an atmospheric residuum could leave tarry materials within the void space of the catalyst, thereby leading to the eventual denaturing of the catalyst with a concurrent loss in catalyst efficiency. Finally, another cause of bed plugging may be due to the physical nature of the catalyst itself—agglomeration of catalyst particles may lead to the formation of catalyst "lumps" within the bed which again leads to catalyst inefficiency, presumably through formation of localized hot spots within the bed. The resulting end point of any of the above deficiencies will be an increased pressure drop through the catalyst bed which eventually requires process shutdown.

Catalyst bed plugging (and a subsequent expensive shutdown) may be minimized by using corrosion-resistant alloys throughout the entire system of feed lines, heaters, reactors, etc., thereby reducing the risk of rust accumulation in the feedstock. Removal of debris that has accumulated in the feedstock from the transportation and storage aspects may be achieved by mechanical filtering. Every attempt should be made to desalt the crude oil to remove any inorganic materials (sodium chloride, etc.) that may originate from the brines that are associated with crude oils as they are pumped to the surface of the earth. In the processing of the heavier feedstocks, it may be advisable to suspend the top layer of catalyst in basket-type devices so that this layer, when plugged, can be conveniently removed and replaced. It may be necessary to use two or more reactors in parallel rather than one large reactor—this would not only lead to more catalyst surface area for collecting debris, but if shutdown was inevitable, one of the reactors could be left onstream while the other reactor was being cleaned. Alternatively, upflow reactors would create the tendency for any debris in the feedstock to fall out of the catalyst, and a circulating stream of liquid below the catalyst bed could be channeled through a filter to remove any solid materials. In a similar manner, prefiltering of the feedstock through a bed of bauxite or similar material will facilitate removal of debris and, in the presence of hydrogen, may also encourage removal of vanadium.

In summary, protection of the catalyst and the reactor is essential if shutdowns are to be minimized, and this is particularly the case when the heavier feedstocks are used in the hydrodesulfurization process. Some measure of protection from the asphaltic constituents can be achieved by removal of these materials by means of a deasphalting step (see Chapter 4), but the overall relative merits of any particular total processing scheme should be assessed in terms of feedstock properties, product yields, process variables, and last but not least, process economics.

5-6 PROCESS VARIABLES

The efficiency of the hydrodesulfurization process is measured by the degree of sulfur removal or, in other words, by the yields of sulfur-free products, but there are several process variables that need special attention as any one of these variables can have a marked influence on the course and efficiency of the hydrodesulfurization process. The major process variables are (1) reactor temperature; (2) hydrogen pressure; (3) liquid hourly space velocity; and (4) hydrogen recycle rate. Other variables such as reactor type and catalyst type have been discussed in an earlier part of this chapter, while the influence of the feedstock type will be discussed in a later chapter.

A. Reactor Temperature

The temperature in the hydrodesulfurization reactor is often considered to be the primary means by which the process is controlled. For example, at stabilized reactor conditions, a rise of 10°C (18°F) in the reaction temperature will substantially increase, and may even double, the reaction rate. Although the data for residua and heavy oil processing are not readily available, an increase in the temperature (from 360 to 380°C, i.e., from 680 to 715°F) for the hydrodesulfurization of a heavy gas oil will increase the conversion slightly (Fig. 5-10) or for a fixed conversion of about 90% enables the quantity of catalyst necessary for the process to be halved. There are, however, limits to which the temperature can be increased without adversely affecting process efficiency; at temperatures above 410°C (770° F), thermal cracking of the hydrocarbon constituents becomes the predominant process which can lead to formation of considerable amounts of low-molecular-weight hydrocarbon liquids and gases. In addition, these high-temperature cracking reactions cannot be diminished by increasing the partial pressure of the hydrogen, and excessively high temperatures (above 400°C or 750°F) lead to deactivation of the catalyst much more quickly than lower temperatures.

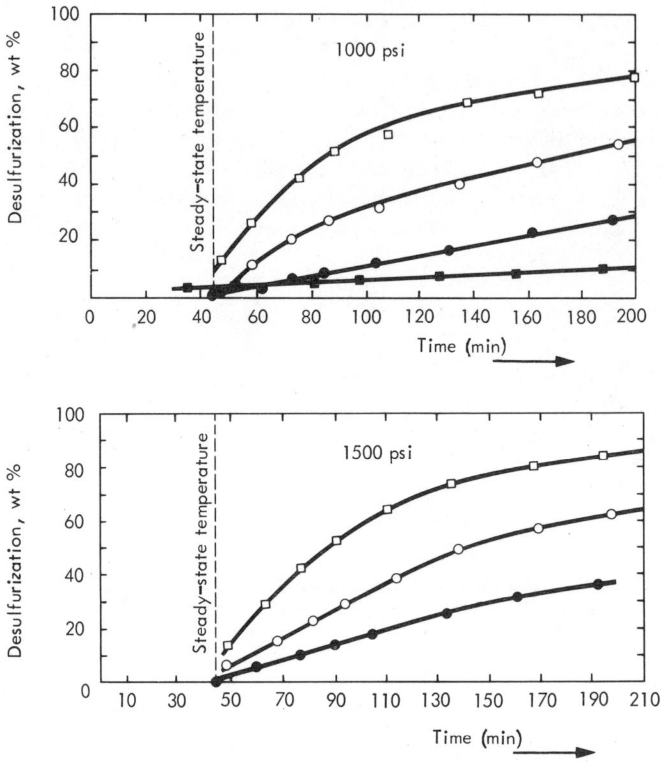

FIGURE 5-10 Variation of desulfurization with temperature for Boscan deasphalted oil (a) at 1000 psi and (b) at 1500 psi. □ = 400 C, ○ = 375 C, ● = 350 C, ■ = 325 C. (From J. Pachano, J. H. Krasuk, O. Rodriguez, and A. Garcia, Proc. Oil Sands of Canada—Venezuela, Can. Inst. Mining, & Metal. Special Vol. No. 17, 1977, p. 219.)

B. Hydrogen Pressure

The overall effect of increasing the partial pressure of the hydrogen is to increase the extent of the conversion through an increase in catalyst activity (Figs. 5-11, 5-12). This is to be expected since the essential function of the catalyst is to serve as a means by which the reactants are brought together, thereby promoting interaction between the feedstock constituents and the hydrogen. However, as with the temperature variable, there are also limitations to increasing the partial pressure of the hydrogen; use of excessively high partial pressures (for conventional feedstocks—above 1000 psi; for heavy feedstocks this figure could be as high

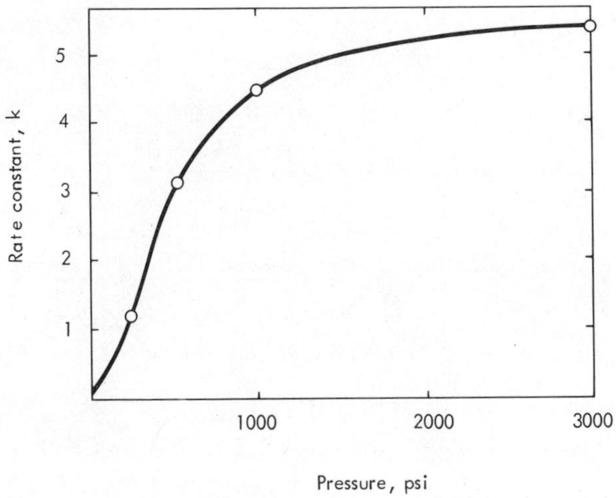

FIGURE 5-11 Effect of pressure on the rate constant for the hydrodesulfurization of a Kuwait atmospheric residuum (36% by volume of crude).

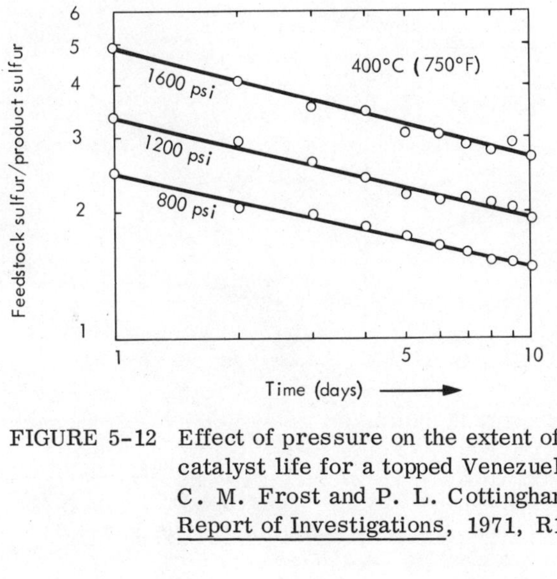

FIGURE 5-12 Effect of pressure on the extent of desulfurization and catalyst life for a topped Venezuelan crude oil. (From C. M. Frost and P. L. Cottingham, Bureau of Mines Report of Investigations, 1971, R1 7557, p. 13.)

as 2000 psi) may only serve to saturate the catalyst, and any increase in the partial pressure of the hydrogen will affect the conversion only slightly.

There are, however, two other instances where the partial pressure of the hydrogen will influence the course of the reaction. The nature of the hydrodesulfurization process is such that the rate will increase (subject to the above limitations) with increasing partial pressure of the hydrogen. However, feedstock conversion decreases with increasing ammonia (a product of the reaction of hydrogen with nitrogen-containing organic constituents of the feedstock) partial pressure and care should be taken to ensure that an increase in the hydrogen partial pressure does not adversely affect the reaction by maintaining a relatively high concentration of ammonia in the vicinity of the catalyst.

In a similar manner, care should be taken that high concentrations of hydrogen sulfide are not allowed to build up at any point in the reactor. Initially, hydrogen sulfide produced by the hydrodesulfurization process may be used beneficially, i.e., to sulfide the catalyst, but high concentrations of this gas will only serve to cause corrosion of the equipment as well as adversely affecting the activity of the catalyst.

C. Liquid Hourly Space Velocity

The liquid hourly space velocity is the ratio of the hourly volume flow of liquid in, say, barrels to the catalyst volume in barrels, and the reciprocal of the liquid hourly space velocity gives the contact time. Since the catalyst volume for the process will be constant, the space velocity will

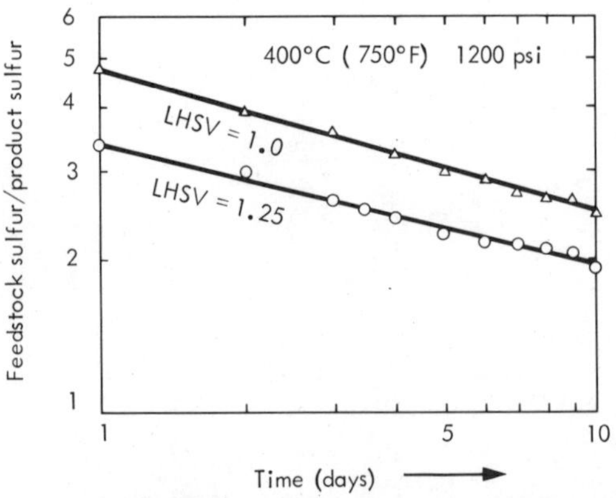

FIGURE 5-13 Effect of space velocity on the rate of desulfurization and catalyst life for a topped Venezuelan crude oil. (From C. M. Frost and P. L. Cottingham, <u>Bureau of Mines Report of Investigations</u>, 1971, R1 7557, p. 12.)

vary directly with the feed rate. A decrease in the liquid hourly space velocity (or, alternatively, an increase in the contact time) will usually bring about an increase in the efficiency (or extent) of the hydrodesulfurization process (Fig. 5-13). In order to maintain a fixed rate of hydrodesulfurization when the feed rate is increased, it may be necessary to increase the temperature.

Since the liquid hourly space velocity is a ratio which involves the use of the volume of catalyst in the reactor, it is also possible to use the data to estimate the additional catalyst required to increase the efficiency of the process. This is especially relevant when the activity of different catalysts is to be compared, and such comparisons are only meaningful if estimations can be made of the quantities of catalyst required to obtain a predetermined performance. For example, to intimate that one catalyst is twice as active as another catalyst signifies that one unit of the first catalyst will function as efficiently as two units of the second catalyst. Such differences in catalyst performance can, however, only be determined accurately through estimations of the liquid hourly space velocity and how this parameter varies with the degree of hydrodesulfurization.

D. Hydrogen Recycle Rate

The optimum use of hydrodesulfurization catalysts requires a relatively high hydrogen partial pressure, and it is therefore necessary to introduce with the feedstock quantities of hydrogen that are considerably greater than required on the basis of stoichiometric chemical consumption. In all cases, process economics dictate that unused hydrogen should be recycled after it has been partially (or completely) freed from hydrogen sulfide that was produced in the previous pass. The overall hydrogen consumption is a summation of several processes which are (1) removal of the sulfur, nitrogen, and oxygen in the feedstock as their hydrogenated analogs, i.e., hydrogen sulfide, ammonia, and water, respectively; (2) addition of hydrogen to unsaturated (olefinic) functions in the products as will occur in the prevailing conditions of the hydrodesulfurization process; (3) destruction, by saturation (i.e., addition of hydrogen), of certain aromatic compound types; and (4) stabilization of unsaturated short-lived organic intermediates that exist during hydrocracking.

Thus, if the hydrogen is not recycled, the process economics are unfavorable and, in addition, the efficiency of the hydrodesulfurization reaction may be adversely affected because of the possible competing reactions outlined above.

BIBLIOGRAPHY

Beuther, H., and Schmid, B. K.: Reaction mechanisms and rates in residue hydrodesulphurisation, Proc. Sixth World Petrol. Cong., 3:297, 1963.

Bland, William F., and Davidson, Robert F. (eds.): Petroleum Processing Handbook, McGraw-Hill, New York, 1967.

Considine, Douglas M. (ed.-in-chief): Energy Technology Handbook, McGraw-Hill, New York, 1977.

Cuenca, A., Galiasso, R., and Andreu, P.: Development of hydrodesulphurisation and hydrodemetallisation catalysts for treating heavy crude oils, Proc. Pacific Chem. Eng. Cong., 2(1):562, 1977.

Frost, C. M., and Cottingham, P. L.: Hydrodesulphurisation of Venezuelan residual fuel oils, U.S. Bureau of Mines, Report of Investigations, RI 7557, 1971.

Hobson, G. D., and Pohl, W. (eds.): Modern Petroleum Technology, Applied Science Publishers, Barking, Essex, England, 1973.

McKinley, Joseph B.: The hydrodesulphurisation of liquid petroleum fractions, Catalysis, 5:405, 1957.

Nelson, Wilbur L.: Hydrogen needed for residual desulphurising, Oil and Gas Journal, 79, December 1, 1969.

Nelson, Wilbur L.: Catalyst consumption required in desulphurising residua, Oil and Gas Journal, 72, November 15, 1976.

Nelson, Wilbur L.: Hydrogen costs run high when desulphurising residua, Oil and Gas Journal, 66, February 21, 1977.

Nelson, Wilbur L.: Data correlation shows the amount of hydrogen used in desulphurising residua, Oil and Gas Journal, 126, February 28, 1977.

Ozaki, H., Satomi, Y., and Hisamitsu, T.: The kinetics and economics of hydrodesulphurising residual fuel oils, Proc. Ninth World Petrol. Cong., 6:97, 1975.

Pachano, J., Krasuk, J. H., Rodriguez, O., and Garcia, A.: Hydrodesulphurisation and hydrodemetalation of a deasphalted oil of Boscan crude with a Ni-W catalyst, Proc. Oil Sands of Canada-Venezuela, Canadian Inst. Mining and Metallurgy, 216, 1977.

Ranney, Maurice W.: Desulphurisation of Petroleum, Noyes Data Corporation, Park Ridge, N.J., 1975.

Richardson, R. L., and Alley, S. K.: Consideration of catalyst pore structure and asphaltenic sulphur in the desulphurisation of resids, in Hydrocracking and Hydrotreating, Am. Chem. Soc. Symposium Series No. 20, 136, 1975.

van Zoonen, D., and Douwes, C. Th.: Effect of pellet pore structure on catalyst performance in the hydrodesulphurisation of straight-run gas oil, J. Inst. Petrol., 49:383, 1963.

6

INFLUENCE OF FEEDSTOCK ON THE HYDRODESULFURIZATION PROCESS

6-1 FEEDSTOCK TYPE

The hydrodesulfurization process variables (Chapter 5) usually require some modification to accommodate the various feedstocks that are submitted for this particular aspect of refinery processing. Although the main purpose of the text is to outline the hydrodesulfurization process with particular reference to the heavier oils and residua, some reference to the lighter feedstocks is warranted which, hopefully, will serve as a base point to indicate the necessary requirements for heavy oil and residuum hydrodesulfurization (Table 6-1).

One particular aspect of the hydrodesulfurization process that needs careful monitoring, with respect to feedstock type, is the exothermic nature of the reaction. The heat of the reaction is proportional to the hydrogen consumption and with the more saturated lower-boiling feedstocks where hydrocracking may be virtually eliminated, the overall heat production during the reaction may be small, leading to a more controllable temperature profile. However, with the heavier feedstocks where hydrogen consumption is appreciable (either by virtue of the hydrocracking that is necessary to produce a usable product or by virtue of the extensive hydrodesulfurization that must occur), it may be desirable to provide internal cooling of the reactor by introducing cold recycle gas to the catalyst bed to compensate for excessive heat. One other generalization that may apply to the lower-boiling feedstocks is that the hydrodesulfurization process may actually have very little effect on the properties of the feedstock (assuming that hydrocracking reactions are negligible)—removal of sulfur will cause some drop in specific gravity which could give rise to volume recoveries approaching (or even above) 100%. Furthermore, with the assumption that cracking reactions are minimal, there may be a slight lowering of the boiling range due to sulfur removal from the feedstock constituents; however, the production of lighter fractions is usually small and may only amount to

TABLE 6-1 Effect of Feedstock Type on Desulfurization Conditions

			Process condition							
	Boiling range		Temperature		Hydrogen pressure		Hydrogen rate		Catalyst life	
Feedstock	°C	°F	°C	°F	psi	kg cm^{-2}	(scf/bbl)	LHSV	Months	bbl/lb
Naphtha	70–170	160–340	300–370	570–700	100–450	7–31.5	250–1500	5–8	36–48	500–1200
Kerosene	160–240	320–465	330–370	625–700	150–500	10.5–35	500–1500	4–6	36–48	300–600
Gas oil	240–350	465–660	340–400	645–750	150–700	10.5–49	1000–2000	2–6	36–48	200–400
Vacuum gas oil	350–650	660–1200	360–400	680–750	450–800	31.5–56	1000–4000	1–3	36–48	50–350
Residua	>650	>1200	370–450	700–840	750–2250	52.5–157.5	1500–10,000	0.5–2	12–24	2–50

some 1 to 5% by weight of the products boiling below the initial boiling point of the feedstock.

A. Light and Middle Distillates

The hydrodesulfurization of light (low-boiling) distillates (naphthas) is one of the more common catalytic hydrodesulfurization processes since it is usually used as a pretreatment of such feedstocks prior to catalytic reforming. Hydrodesulfurization of such feedstocks is required because sulfur compounds poison the precious-metal catalysts used in reforming and desulfurization can be achieved under relatively mild conditions and is near-quantitative (Tables 6-2, 6-3). If the feedstock arises from a cracking operation, hydrodesulfurization will be accompanied by some degree of saturation resulting in increased hydrogen consumption.

The hydrodesulfurization of low-boiling (naphtha) feedstocks is usually a gas-phase reaction and may employ the catalyst in fixed beds and (with all of the reactants in the gaseous phase) only minimal diffusion problems are encountered within the catalyst pore system. It is, however, important that the feedstock be completely gasified before entering the reactor as there may be the possibility of pressure variations (leading to less satisfactory results) if some of the feedstock enters the reactor in the liquid phase and is vaporized within the reactor.

In applications of this type, the sulfur content of the feedstock may vary from 100 ppm to 1% and the necessary degree of desulfurization to be effected by the treatment may vary from as little as 50% to more than 99%. If the sulfur content of the feedstock is particularly low, it will be necessary to presulfide the catalyst. For example, if the feedstock only has 100 to 200 ppm sulfur, several days may be required to sulfide the catalyst as an integral part of the desulfurization process even with complete reaction of all of the feedstock sulfur to, say, cobalt and molybdenum (catalyst) sulfides. In such a case, presulfiding can be conveniently achieved by addition of sulfur compounds to the feedstock or by addition of hydrogen sulfide to the hydrogen.

Generally, hydrodesulfurization of naphtha feedstocks to produce catalytic reforming feedstocks is carried to the point where the desulfurized feedstock contains less than 20 ppm sulfur. The net hydrogen produced by the reforming operation may actually be sufficient to provide the hydrogen consumed in the desulfurization process.

The hydrodesulfurization of middle distillates is also an efficient process (Tables 6-4 and 6-5) and applications include predominantly the desulfurization of kerosene, diesel fuel, jet fuel, and heating oils which boil over the general range 250 to 400°C (480 to 750°F). However, with this type of feedstock, hydrogenation of the higher-boiling catalytic cracking feedstocks has become increasingly important where hydrodesulfurization is accomplished alongside the saturation of condensed-ring aromatic compounds as an aid to subsequent processing.

Under the relatively mild processing conditions used for the hydrodesulfurization of these particular feedstocks, it is difficult to achieve complete vaporization of the feed, and process conditions may dictate that only part of the feedstock is actually in the vapor phase and that sufficient liquid phase is maintained in the catalyst bed to carry the larger molecular constituents of the feedstock through the bed. If the amount of liquid phase is insufficient for this purpose, molecular stagnation (leading to carbon deposition on the catalyst) will occur.

TABLE 6-2 Hydrodesulfurization of a Low-Boiling Distillate (Naphtha)

Properties of feedstock:	
Specific gravity	0.743
°API	59.0
ASTM distillation, °C	
IBP	85
10% vol	107
50% vol	130
90% vol	159
EP	172
Sulfur, wt %	0.04
Process conditions:	
Catalyst	Cobalt molybdate on alumina
Temperature, °C	260-370
Pressure, psi	200-500
Properties of product:	
Specific gravity	0.740
°API	59.7
ASTM distillation, °C	
IBP	71
10% vol	106.5
50% vol	133
90% vol	160.5
EP	175
Sulfur, wt %	0.0003

TABLE 6-3 Hydrodesulfurization of Cracked Naphthas of Varying Origins

Feed	Boiling range		Sulfur (wt %)	Desulfurization (%)
	°F	°C		
Visbreaker naphtha	150–430	65–220	1.00	90
Visbreaker/coker naphtha	150–430	65–220	1.03	85
Heavy catalytic naphtha	250–440	120–225	0.24	71
Heavy thermal naphtha	300–450	150–230	0.28	57
Light catalytic naphtha	200–350	95–175	0.18	89

TABLE 6-4 Hydrodesulfurization of a Straight-Run Middle Distillate

Feedstock	
Specific gravity	0.844
°API	36.2
ASTM distillation, °C	
IBP	238
10% vol	265
50% vol	288
90% vol	332
EP	368
Sulfur, wt %	1.20
Aniline point, °C	76.0
Pour point, °C	-7
Conradson carbon (10% residuum)	0.024
Viscosity, cSt (38°C)	4.14
Process conditions:	
Catalyst	Cobalt molybdate on alumina
Temperature, °C	315-430
Pressure, psi	250-1000
Product	
Specific gravity	0.832
°API	38.6
ASTM distillation, °C	
IBP	222
10% vol	259
50% vol	287
90% vol	330
EP	336
Sulfur, wt %	0.09
Aniline point, °C	79.5
Pour point, °C	-6
Conradson carbon (10% residuum)	0.022
Viscosity, cSt (38°C)	3.77

TABLE 6-5 Hydrodesulfurization of a "Cracked" Middle Distillate

Feedstock	
Specific gravity	0.901
°API	25.5
ASTM distillation, °C	
IBP	227
10% vol	242
50% vol	256
90% vol	277
EP	293
Sulfur, wt %	2.34
Aniline point, °C	32.4
Pour point, °C	-28
Conradson carbon (10% residuum)	0.34
Viscosity, cSt (38°C)	2.38
Process conditions:	
Catalyst	Cobalt molybdate on alumina type
Temperature, °C	315-430
Pressure, psi	250-1000
Product	
Specific gravity	0.876
°API	30.0
ASTM distillation, °C	
IBP	203
10% vol	227
50% vol	250
90% vol	275
EP	293
Sulfur, wt %	0.35
Aniline point, °C	36.0
Pour point, °C	-28
Conradson carbon (10% residuum)	0.05
Viscosity, cSt (38 °C)	2.22

The hydrodesulfurization of the middle distillates causes a more marked change in the specific gravity of the feedstock, and the amount of low-boiling material is much more significant when compared with the naphtha-type feedstock. In addition, the somewhat more severe reaction conditions (leading to some degree of hydrocracking) also leads to an overall increase in hydrogen consumption when middle distillates are employed as feedstocks in place of the naphthas.

B. High-Boiling Distillates

The high-boiling distillates, such as the atmospheric and vacuum gas oils, are not usually produced as a refinery product but merely serve as feedstocks to other processes for conversion to lower-boiling materials. For example, gas oils can be desulfurized to remove more than 80% of the sulfur originally in the gas oil with some conversion of the gas oil to lower-boiling materials (Table 6-6). The treated gas oil (which has a reduced carbon residue as well as lower sulfur and nitrogen contents relative to the untreated material) can then be converted to lower-boiling products in, say, a catalytic cracker where an improved catalyst life and volumetric yield may be noted.

The conditions used for the hydrodesulfurization of a gas oil may be somewhat more severe than the conditions employed for the hydrodesulfurization of middle distillates with, of course, the feedstock in the liquid phase.

In summary, the hydrodesulfurization of the low-, middle-, and high-boiling distillates can be achieved quite conveniently using a variety of processes. One major advantage of this type of feedstock is that the catalyst does not become poisoned by metal contaminants in the feedstock as (at most) only negligible amounts of these contaminants will be present. Thus, the catalyst may be regenerated several times and onstream times between catalyst regeneration (while varying with the process conditions and application) may be of the order of 3 to 4 years (Table 6-1).

C. Heavy Oils and Residua

Heavy oils and residua usually require more severe hydrodesulfurization conditions to produce low-sulfur liquid product streams which can then, as is often now desired, be employed as feedstocks for other refining operations. Hydrodesulfurization of the heavier feedstocks is normally accompanied by a high degree of hydrocracking, and thus the process conditions required to achieve 70 to 90% desulfurization will also effect substantial conversion of the feedstock to lower-boiling products. The extent of the hydrocracking is, like the hydrodesulfurization reaction, dependent upon the temperature, and both reaction rates increase with increase in temperature. However, the rate of hydrocracking tends to show more marked increases

TABLE 6-6 Hydrodesulfurization of High-Boiling Distillates

Feedstock		
Source	Kuwait	Khafji
Boiling range, °C	370-595	370-595
Specific gravity	0.935	0.929
°API	19.8	20.2
Sulfur, wt %	3.25	3.05
Process conditions:		
Temperature, °C	370-430	
Pressure, psi	100-500	
Hydrogen consumption, scf/bbl	420	400
Product		
Naphtha		
C_5-205°C		
Yield, vol %	1.7	1.7
Specific gravity	0.802	0.802
°API	45.0	45.0
Sulfur, wt %	0.02	0.02
Higher boilers		
(>205°C)		
Yield, vol %	99.6	99.5
Specific gravity	0.897	0.893
°API	26.3	27.0
Sulfur, wt %	0.50	0.48

with temperature than the rate of hydrodesulfurization. The overall effect of the increase in the rate of the hydrocracking reaction is to increase the rate of carbon deposition on the catalyst which adversely affects the rate of hydrodesulfurization; hydrocracking reactions are not usually affected by carbon deposition on the catalyst since they are more dependent upon the noncatalytic scission of covalent bonds brought about by the applied thermal energy.

In contrast to the lighter feedstocks that may be subjected to the hydrodesulfurization operation, the heavy oils and residua may need some degree of pretreatment. For example, the process catalysts are usually susceptible to poisoning by nitrogen (and oxygen) compounds and metallic salts (in addition to the various sulfur-compound types) which tend to be concentrated in residua (Chapter 1) or exist as an integral part of the heavy oil matrix.

Thus, any processing sequence devised to hydrodesulfurize a heavy oil or residuum must be capable of accommodating the constituents which adversely affect the ability of the catalyst to function in the most efficient manner possible.

The conditions employed for the hydrodesulfurization of the heavier feedstocks may be similar to those applied to the hydrodesulfurization of gas oil fractions but with the tendency to increased pressures. However, carbon deposition on, and metal contamination of, the catalyst is much greater when residua and heavy oils are employed as feedstocks and, unless a low level of desulfurization is acceptable, frequent catalyst regeneration is necessary.

A wide choice of commercial processes are available for the catalytic hydrodesulfurization of heavy oils and residua (Chapter 7). The suitability of any particular process depends not only upon the nature of the feedstock but also on the degree of desulfurization that is required as well as on the relative amounts of the lower-boiling products that are to be produced as feedstocks for further refining and generation of liquid fuels.

There is, however, one aspect of feedstock properties that has not yet been discussed fully and that is feedstock composition. This particular aspect of the nature of the feedstock is, in fact, related to the previous section where the influence of various feedstock types on the hydrodesulfurization process was noted, but it is especially relevant when residua and heavy oils from various sources are to be desulfurized.

6-2 FEEDSTOCK COMPOSITION

The composition of the various feedstocks may, at first sight, seem to be of minor importance when the problem of the hydrodesulfurization of heavy oils and residua comes under consideration. However, consideration of the variation in process conditions that were outlined in the previous section (Table 6-1) for different feedstocks (where the feedstocks are relatively well-defined boiling fractions of petroleum) presents some indication of the problems that may be encountered where the feedstocks are less well defined not only from the point of view of their molecular weight ranges but also from their molecular composition. Such is the nature of the problem when dealing with various residua and heavy oils which are (to say the least) unknown in terms of their chemical composition. In fact, the complexity of these particular materials (Chapter 1) has allowed little more than speculation as to the molecular structure of the constituents.

Nevertheless, there have been many successful attempts to define the behavior of residua (and heavy oils) during the hydrodesulfurization process in terms of physical composition, which has also led to the development of process modifications to suit various heavy feedstocks. This line of investigation arose because of the tendency, over the years, to classify all residua (and heavy oils) as the same type of refinery discard which was, effectively, useless for further processing—the only exception being the production of asphalts from certain residua. It was only when these so-called useless materials entered the refinery operation as feedstocks for the production of additional liquid fuels that it was realized that each particular residuum or heavy oil may have to be considered for processing on the basis of an examination of several fundamental properties.

The means by which heavy oils and residua can be evaluated has been described in an earlier chapter (Chapter 2) and it is therefore the purpose of the present section to relate the physical properties (or evaluation) of these feedstocks to their performance in the hydrodesulfurization process with special reference to any parameters that adversely affect process efficiency.

A. Asphaltics Content

Residua and heavy oils, like any other petroleum, can be fractionated by a variety of techniques to provide broad general fractions termed asphaltenes, resins, and oils. Briefly, the asphaltenes fraction of a petroleum is that fraction which is precipitated by the addition of a large excess of a low-boiling liquid hydrocarbon (usually n-pentane). On the other hand, resins are those materials which remain soluble in the pentane but will be adsorbed by a surface-active material such as fuller's earth, while the oils fraction is soluble in pentane but is not adsorbed from the pentane solution by any surface-active material. The asphaltics fraction of any petroleum, heavy oil, or residuum is usually a combination of the asphaltene and resin fraction and, in many instances, may constitute a large portion of a heavy oil or, especially, of a residuum.

The asphaltic fraction of any petroleum type contains the higher-molecular-weight materials and contains higher proportions of sulfur, nitrogen, and oxygen than the oils fraction. The H/C atomic ratios will vary from about 1.0 in the asphaltenes to about 1.3 in the resins, indicating significant amounts of aromatic centers, while the more aliphatic oils fraction will have an atomic H/C ratio of the order of 1.6. In any investigations on the asphaltic constituents of residua, attention has usually been devoted to the asphaltene fraction, with the resin fraction receiving considerably less attention. Nevertheless, the resin fraction of residua and heavy oils is just as important as the asphaltene fraction since the resins have the potential to be converted into asphaltenes under remarkably mild conditions either through the application of heat or through exposure to aerial oxygen. Therefore, in any assessment of the influence of the asphaltene fraction on the processability of a residuum (or similar material), attention should also be paid to the proportion of the resins in that material because of the potential of this fraction to act as a precursor to the asphaltenes and, eventually, to coke.

With respect to the influence of the asphaltics fraction on the processability of a heavy feedstock, a high asphaltics content will require high hydrogen partial pressures and may actually limit the maximum level of hydrodesulfurization, or final traces of sulfur in the residuum may only be eliminated under extremely severe reaction conditions where hydrocracking is the predominant reaction in the process. High asphaltics contents of feedstocks are also responsible for high viscosities (Fig. 6-1) which may increase the resistance to mass transfer of the reactants to the catalyst surface and thus decrease the rate of sulfur removal. In an instance such as this, it may be deemed satisfactory by the refiner to simply blend the heavy feedstock with a lighter feedstock, thereby alleviating the problem for that particular process. Similarly, a high asphaltics content of a feedstock usually indicates a high sulfur content (Fig. 6-2) of that particular

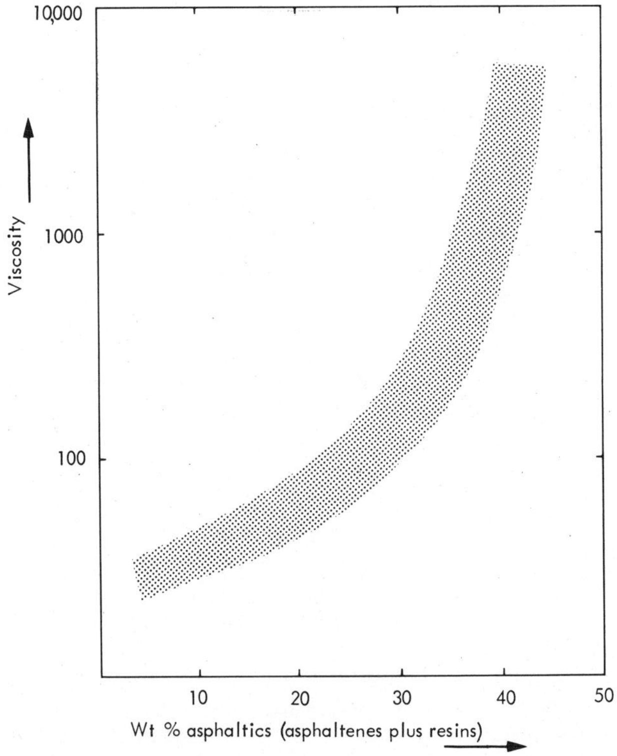

FIGURE 6-1 The variation of viscosity with the asphaltics (asphaltenes plus resins) content of crude oil.

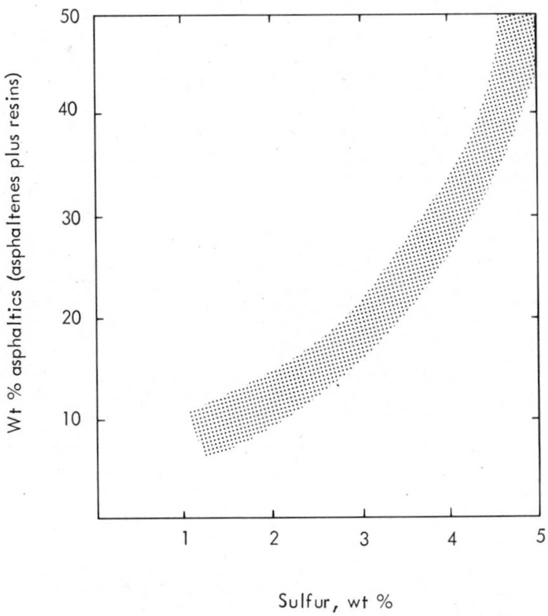

FIGURE 6-2 Variation of sulfur content with the asphaltics (asphaltenes plus resins) content of crude oils.

feedstock, which will require the higher partial pressure of hydrogen as well as the more severe reaction conditions indicated in the earlier part of this section. The sulfur content of a feedstock may also be expressed in terms of the asphaltene content alone (Fig. 6-3), there seemingly being a relationship between the asphaltenes and the asphaltene sulfur as well as (for undefined reasons) a relationship between the asphaltenes and nonasphaltene sulfur (Fig. 6-4).

It is generally true that the sulfur content of a residuum or heavy oil has a marked influence on the characteristics of the desulfurization process. Thus, for a series of residual feedstocks (Table 6-7) the relative rate constants for the desulfurization process decrease with increasing sulfur content (Fig. 6-5) under standard process conditions. With the general proviso that the majority of the sulfur in a feedstock is located in the asphaltics fraction (i.e., the asphaltenes plus the resins), it is not surprising that there have been many suggestions to improve the desulfurization process by prior removal of these extremely complex fractions. There are also other problems that arise by virtue of the presence of these fractions in a feedstock; for example, the asphaltic constituents of feedstocks also pose the

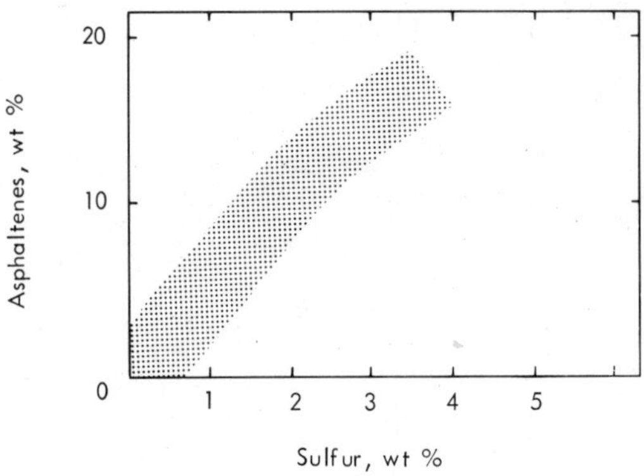

FIGURE 6-3 Variation of sulfur content with asphaltene content of crude oils.

problem of coke deposition. The coke-forming tendencies of most feedstocks can be directly estimated from (and are, in fact, related to) the carbon residue of that feedstock for that particular process, and the higher the asphaltics content of a feedstock, the higher the carbon residue (Fig. 6-6). It has been postulated that the asphaltene fraction is the sole

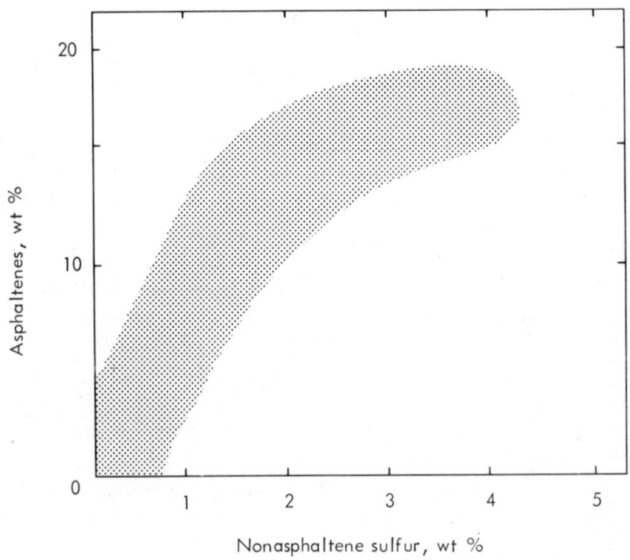

FIGURE 6-4 Variation of asphaltene content with nonasphaltene sulfur.

TABLE 6-7 Properties of Various Residua

Crude source	Vol % of crude		Properties of 650 + residuum			
	650°F + residue vol % crude	1050°F + residue vol % crude	API gravity	Sulfur (wt %)	Metals (ppm) (Ni + V)	Carbon residue (Conradson) (wt %)
Murban	35.7	9.1	22.8	1.5	4	3.6
North Slope	52.6	18	15.1	1.74	45	8.9
Agha Jari (Iranian light)	42.7	13.2	17.5	2.37	98	7.7
Venezuelan (Leona)	55.0	10	18.9	2.1	187	8.6
Arabian light	44.7	12.7	17.9	2.89	28	7.3
L.A. Basin (Torrey Canyon)	48.4	36	8.4	1.97	236	11.7
West Texas Sour	41.6	12.1	15.5	3.38	29	9.0
Gach Saran (Iranian heavy)	46.7	19.1	14.6	2.55	258	9.9
Sassan	44.7	14.0	16.0	3.42	38	8.5
Kirkuk	39.3	12.6	17.6	3.7	84	8.4
Kuwait	50.5	19.8	15.5	3.9	51	8.8
Safaniyah (Arabian heavy)	53.1	26.8	12.3	4.19	123	13.3
Khafji	53.0	21.0	11.9	4.3	75	15.6
Ratawi	62.0	40	10.2	5.0	57	14.1

Source: Reproduced from M. Skripek, S. K. Alley, P. F. Helfrey, and P. F. Leuth. Proceedings of the 79th National Meeting of the American Institute of Chemical Engineers, Houston, Texas, March 1978; with permission of the publisher.

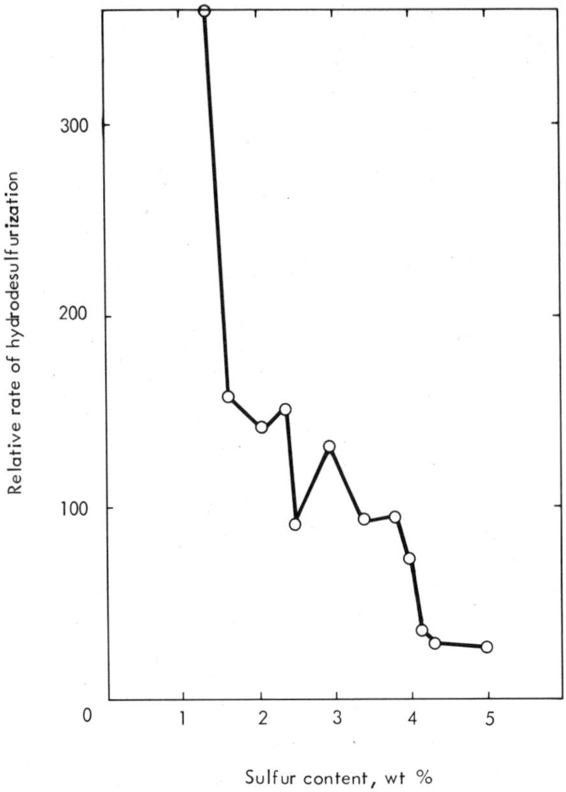

FIGURE 6-5 Variation in the relative hydrodesulfurization rate with sulfur content of the feedstock. (Reproduced from M. Skripek, S. K. Alley, P. F. Helfrey, and P. F. Leuth. Proceedings of the 79th National Meeting of the American Institute of Chemical Engineers, Houston, Texas, March 1978, with permission of the publisher).

source of coke deposited during processing various feedstocks, but there is some evidence to indicate that not only the resins but also some of the heavier constituents of the gas oil also contribute to coke formation. However, coke formation from the asphaltenes is presumed to be the more rapid reaction rate. Again, as with the problem of sulfur removal from the asphaltics constituents, some alleviation of the problem of coke formation can be achieved by increasing the partial pressure of the hydrogen in the process (Fig. 6-7).

One suggested method of facilitating hydrodesulfurization of heavy oils and residua is to remove completely, or partially, all of the asphaltics from the stock. In the first instance, the asphaltic constituents can be completely, and conveniently, removed from a heavy stock by treatment with liquid propane under pressure whereupon the asphaltenes and the resins are deposited as a heavy, thick, black asphalt leaving the lower-molecular-weight oils fraction in the liquid phase. Alternatively, use of other hydrocarbon solvents allows adjustment of the yield of precipitate in the form of partial removal of the asphaltics (Fig. 6-8), thereby retaining some, or all, of the resins in the stock to serve as potential liquid-forming materials.

Thus, although the concept of deasphalting has been discussed elsewhere as a potential desulfurizing technique (Chapter 4), it has also been advocated as a method of feedstock pretreatment to remove the objectionable asphaltic constituents from feedstocks that are destined for catalytic hydrodesulfurization. For example, removal of part of a Kuwait residuum as asphaltic material (thereby removing macromolecular species containing high proportions of sulfur, nitrogen, and oxygen) allows a severalfold increase in space velocity while maintaining a comparable level of

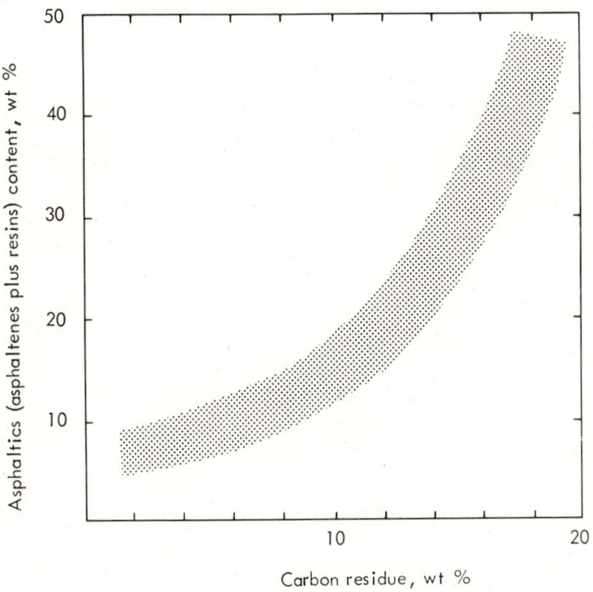

FIGURE 6-6 Variation of carbon residue (Conradson) with the asphaltics (asphaltenes plus resins) content of crude oils.

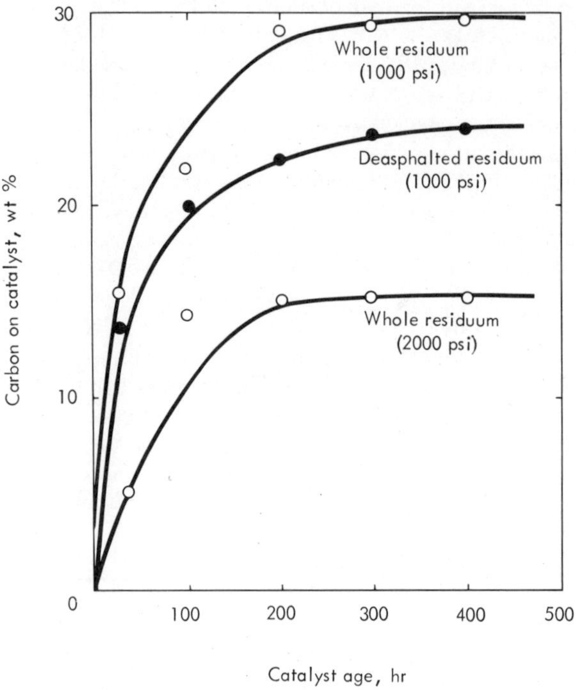

FIGURE 6-7 Carbon deposit as a function of time in the hydrodesulfurization of Kuwait residuum. [From H. Beuther and B. K. Schmid, Proc. 6th World Petrol. Cong., 1963, 3, 305; reproduced with permission.]

desulfurization and substantially improving the lifespan of the hydrodesulfurization catalyst, or it allows substantial increases in the rate of hydrodesulfurization (Table 6-8). However, in an instance such as this, the yield of gasoline may remain the same for the deasphalted material as for the whole residuum (Table 6-9), but the overall quality of the residual fraction is considerably improved when the deasphalted material is used as the feedstock.

In summary, the presence of large proportions of asphaltic materials in hydrodesulfurization feedstocks can adversely affect the process in many ways, not the least of which are reduced catalyst life and hydrogen requirements. It appears that there may be a case for deasphalting the feedstock prior to the hydrodesulfurization step, but obviously each feedstock should be evaluated on its own merits, and the process economics need careful examination before any decision is made on a particular refining sequence.

TABLE 6-8 Hydrodesulfurization of a Kuwait Vacuum Residue and Deasphalted Products

Feedstocks	% S before extraction	% S after extraction	% S after hydrodesulfurization	Hydrodesulfurization rate
Kuwait vacuum residue	5.2	5.2	2.12	60
Pentane-deasphalted oil	5.2	4.3	0.65	85
Butane-deasphalted oil	5.2	3.75	0.3	92
Propane-deasphalted oil	5.2	2.75	0.08	97

TABLE 6-9 Hydrodesulfurization of a Kuwait Residuum and the Deasphalted Residuum (Hydrogen Partial Pressure = 1000 psi)

	Kuwait vacuum residue (18% by vol of crude)	Deasphalted Kuwait vacuum residue (15% by vol of crude)
Yields		
Percent by vol of charge gasoline (IBP-400°F)	5	8
Furnace oil (400-670°F)	22	22
Residue (670°F +)	72	69
Inspection of residue (670°F +)		
°API	12.4	17.7
Sulfur, wt %	1.4	0.5
Nitrogen, wt %	0.42	0.27
Carbon residue, wt % (Conradson)	15.0	7.2
Vanadium, ppm	14	0.2

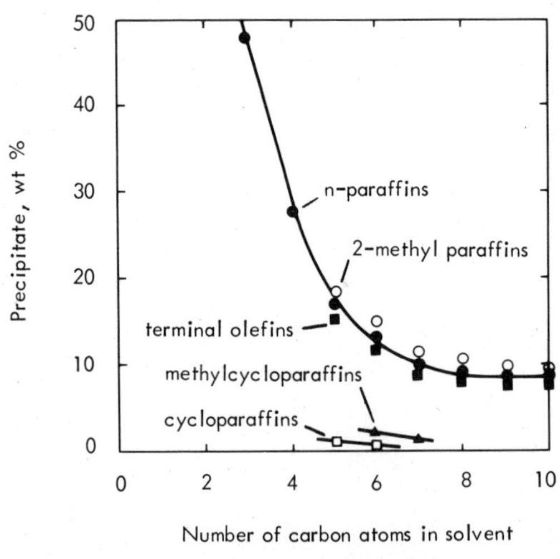

FIGURE 6-8 Relationship of asphaltics removal from a natural bitumen to solvent type and molecular weight. [From D. L. Mitchell and J. G. Speight, Fuel, 1973, 52, 149; reproduced with permission.]

B. Metals Content

Residua and heavy oils contain impurities other than sulfur, nitrogen, and oxygen, and the most troublesome of these impurities are the organometallic compounds of nickel and vanadium. The metal content of a residuum or heavy oil can vary from several parts per million to more than 1000 per million (Table 6-10), and there does seem to be a more than chance relationship between the metals content of a feedstock and its physical properties (Fig. 6-9). In the hydrodesulfurization of the heavier feedstocks the metals (nickel plus vanadium) are an important factor since large amounts (over 150 ppm) will cause rapid deterioration of the catalyst. The free metals, or the sulfides, deposit on the surface of the catalyst and within the pores of the catalyst, thereby poisoning the catalyst by making the more active catalyst sites inaccessible to the feedstock and the hydrogen. This results in frequent replacement of an expensive process commodity unless there are adequate means by which the catalyst can be regenerated.

The problem of metal deposition on the hydrodesulfurization catalysts has generally been tackled by any one of three methods: (1) suppressing deposition of the metals on the catalyst; (2) development of a catalyst which will accept the metals and can tolerate high levels of metals without marked reduction in the hydrodesulfurization capabilities of the catalyst; and (3) removal of the metal contaminants before the hydrodesulfurization step.

The first two methods involve a careful and deliberate choice of the process catalyst as well as the operating conditions and may only be viable

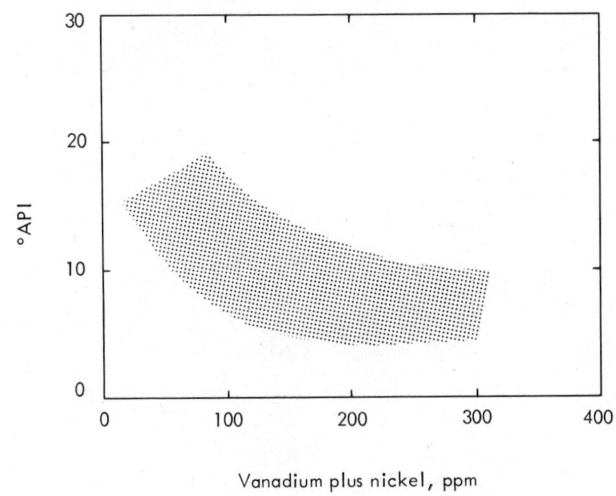

FIGURE 6-9 Variation of metals (vanadium plus nickel) content with the API gravity of various feedstocks (see also Figs. 2-7, 2-8, 2-9, and 2-10).

TABLE 6-10 Metals and Sulfur Content of Various Feedstocks

Name	°API	V + Ni (ppm)	%S
Alaska, Simpson	~4	100-212	2.4-3.1
Alaska, Hurl State	~6	72-164	1.0-1.9
Arabian light, atmospheric	16.1-17.3	34-42	2.8-3.0
Arabian light, vacuum	6.1-8.2	101-112	4.0-4.2
Arabian light, vacuum	11.2	71	3.71
Arabian medium, atmospheric	~15.0	51-140	4.0-4.3
Arabian heavy, atmospheric	11.0-17.2	102-140	4.2-4.6
Safaniya	11.0-12.7	102-130	4.3
Khursaniya	15.1	41	4.0
Khafji, atmospheric	14.7-15.7	95-125	4.0-4.3
Khafji, vacuum	5.0	252	5.4
Wafra, Eocene	9.7	125	5.6
Wafra, Ratawi	14.2	100	4.7-5.2
Agha Jari	16.6	92-110	2.3-2.5
Agha Jari, vacuum	7.5	274	3.8
Iranian light, vacuum	9.5	150-316	3.1-3.2
Iranian heavy, atmospheric	15.0-19.8	179-221	2.5-2.8
Alaska, Kuparup	~11	64-153	1.5-2.4
Alaska, Put River	~10	34-83	1.5-2.1
Alaska, Sag River	~8	52-129	1.6-2.2
California, Coastal	4.3	300	2.3
California, atmospheric	13.7	134	1.73
California, deasphalted	13.8	24	0.95
Delta Louisiana	11.3	25	0.93
Mid-Continent, vacuum	14.5	25	1.2
Wyoming, vacuum	6.3	292	4.56
Salamanca, Mexico	5.4-10.7	182-250	3.2-3.8

TABLE 6-10 (Continued)

Name	°API	V + Ni (ppm)	% S
Gach Saran	15.8–17.0	210–220	2.4–2.6
Gach Saran, vacuum	9.2–10.0	329–359	3.1–3.3
Darius, vacuum	12.8	61	4.7
Sassan, atmospheric	18.3	33	3.3
Kuwait, atmospheric	10.2–16.7	60–65	3.8–4.1
Kuwait, vacuum	5.5–8.3	110–199	5.1–5.8
Zubair, Iraq	18.1	43	3.3
Khurais	14.6	31	3.3
Murban	24.0–25.0	2–10	1.5–1.6
Qatar marine	25.2	35	2.77
Middle East, vacuum	8.3–17.6	40–110	3.0–5.8
Pilon (Tar belt)	9.7	608	3.92
Monogus (Tar belt)	12.0	254	2.5
Morichal (Tar belt)	9.6–12.4	233–468	2.1–4.1
Melones II (Tar belt)	9.9	424	3.3
Tucupita	16.0	129	1.03
Bachaquero (Lake)	5.8–14.6	415–970	2.2–3.5
Boscan (Lake)	9.5–10.4	1185–1750	5.2–5.9
Lagunillas (Lake)	~16.5	271–400	2.2–2.6
Taparita (Lake)	16.9	490	2.4
Tia Juana (Lake)	6.5–7.8	380–705	2.4–3.5
Tia Juana (Lake)	~8.0	327–534	2.6–4.0
Urdaneta	11.7	480	2.68
Venezuela, atmospheric	11.8–17.2	200–460	2.1–2.8
Venezuela, vacuum	4.5–7.5	690–760	2.9–3.2
West Texas, atmospheric	15.4–17.9	39–40	2.5–3.7
West Texas, vacuum	7.3–10.5	65–88	3.2–4.6

TABLE 6-11 Comparative Results for the Desulfurization and Demetallization/Desulfurization of a Venezuelan Residuum

Feedstock data	Type operation: Desulfurization	Type operation: Demetallization/desulfurization
Gravity, °API	11.8	12.6
Sulfur, wt %	2.8	2.8
Vanadium, ppm	375	398
Nickel, ppm	55	57
975°F +, vol %	55	55

Product yield	Desulfurization			Demetallization/desulfurization		
	Yield (vol %)	Sulfur (wt %)	Vanadium (ppm)	Yield (vol %)	Sulfur (wt %)	Vanadium (ppm)
C_4-end point	103.5	0.64	170	103.3	0.64	88
Chemical H_2 consumption, scf/bbl	—	720	—	—	680	—
Desulfurization, %[a]	—	77	—	—	77	—

[a] For two Co/Mo reactors in series.

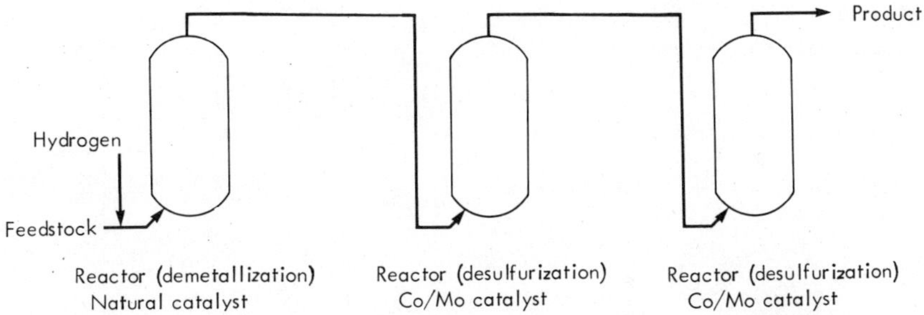

FIGURE 6-10 Schematic representation of a demetallization/hydrodesulfurization process. (From K. E. Hastings, L. C. James, and W. R. Mounce, Oil and Gas Journal, June 30, 1975, p. 128.)

for feedstocks with less than 150 ppm total metals since the decrease in the desulfurizing activity of the catalyst is directly proportional to the metals content of the feedstock. There are, however, catalysts which can reputedly tolerate substantial proportions of metals within their porous structure before the desulfurizing capability drops to an unsatisfactory level. Unfortunately, data on such catalysts are extremely limited because of their proprietary nature, and details are not always available but tolerance levels for metals that are equivalent to 15 to 65% by weight of the catalyst have been quoted.

The third method may be especially applicable to feedstocks with a high metals content and requires a separate demetallization step just prior to the hydrodesulfurization reactor. Such a step might involve passage of the feedstock through a demetallization chamber which contains a catalyst with a high selectivity for metals but whose activity for sulfur removal is low. Nevertheless, demetallization applied as a separate process can be used to generate low-metal feedstocks and will allow a more active, and stable, desulfurization system so that a high degree of desulfurization can be achieved on high-metal feedstocks with an acceptable duration of operation (Fig. 6-10; Table 6-11).

BIBLIOGRAPHY

Billon, A., Lorenz, E., and Peries, J. P.: Valorisation of crude bottoms, Proc. Third International Symposium on Large Chemical Plants, New Technologies, and Varying Feedstocks, Antwerp, 115, October 20-21, 1976.

Billon, A., Peries, J. P., Fehr, E., and Lorenz, E.: SDA key to upgrading heavy crudes, Oil and Gas Journal, 43, January 24, 1977.

Considine, Douglas M. (ed.-in-chief): Energy Technology Handbook, McGraw-Hill, New York, 1977.

Ditman, John G.: Deasphalt to get feed for lubes, Hydrocarbon Processing, 110, May 1973.

Flinn, R. A., Beuther, H., and Schmid, B. K., Now you can improve residue treating, Petroleum Refiner, 40(4):139, 1961.

Frost, C. M., and Cottingham, P. L.: Hydrodesulphurisation of Venezuelan residual fuel oils, U.S. Bureau of Mines, Report of Investigations 7557, 1971.

Furimsky, Edward: Chemical origin of coke deposited on a catalyst surface, Ind. Eng. Chem., Prod. Res. and Devel., 17:329, 1978.

Gleim, W. K. T.: The role of asphaltene in refining, in Developments in Petroleum Science, 7—Bitumens, Asphalts, and Tar Sands (G. V. Chilingarian and T. F. Yen, eds.), Elsevier, Amsterdam, 1978, chap. 10.

Hastings, Kenneth E., James, Lewis C., and Mounce, William R.: Demetallisation cuts desulphurisation costs, Oil and Gas Journal, 126, June 30, 1975.

Hobson, G. D., and Pohl, W. (eds.): Modern Petroleum Technology, Applied Science Publishers, Barking, Essex, England, 1973.

McKinley, Joseph B.: The hydrodesulphurisation of liquid petroleum fractions, in Catalysis, Volume 5, Reinhold, New York, 1957, chap. 6.

Nelson, Wilbur L.: Metals and sulphur affect residua, Oil and Gas Journal, 191, November 8, 1976.

Schuman, S. C., and Shalit, H.: Hydrodesulphurisation, Catalysis Reviews, 4(2):245, 1970.

Skripek, M., Alley, S. K., Helfrey, P. F., and Lueth, P. F.: Importance of feed characteristics in resid desulphurisation, Presented at the 79th National Meeting, American Institute of Chemical Engineers, Houston, Texas, March 16-20, 1975.

Speight, James G.: The Structure of Petroleum Asphaltenes—Current Concepts, Information Series No. 81, Alberta Research Council, Edmonton, Alberta, Canada, 1978.

Viloria, D. A., Krasuk, J. H., Rodriguez, O., Buenafama, H., and Lubkowitz, J.: Solvent deasphalt Boscan crude, Hydrocarbon Processing, 109, March 1977.

7

COMMERCIAL PROCESSES

The technical literature contains many references to processes which are claimed to be suitable for the hydrodesulfurization of heavy oils and residua but some confusion may exist because of the many alternate processes that are available and because of the nomenclature that has been employed by the respective owners of the processes. In essence, all of the hydrodesulfurization processes have one particular aim—to convert a high-sulfur, low-API-gravity, high-viscosity feedstock to a low-sulfur, high-API-gravity, low-viscosity product. In the case of the sulfur, as the name implies, a hydrodesulfurization process removes the sulfur as hydrogen sulfide and, at the same time, hydrogen is added to the parent molecule. However, with respect to the conversion of low-API-gravity (high-viscosity) feedstocks to high-API-gravity (low-viscosity) products, there are other factors involved. Sulfur removal will, to some extent, alter the API gravity (or viscosity) of feedstocks, but major alterations in these particular properties are usually achieved by the use of more severe process conditions, i.e., hydrocracking. It is perhaps in this aspect of hydrodesulfurization that the various processes show greater differentiation.

In this chapter, brief descriptions of the various hydrodesulfurization processes are presented with some attempt to illustrate how the processes vary in their capacity to treat the heavier feedstocks to yield the desired low-sulfur products. There are many processes described in the technical and commercial literature, but a large proportion of these processes (although actively reported) have not yet left the pilot (or exploratory) stage and are, perhaps, a considerable way from commercialization. Thus, the processes described in this chapter are the more common, but commercially proven, processes with the inclusion of some of the more promising processes as well as some that may have just "gone onstream."

7-1 SHELL RESIDUAL OIL HYDRODESULFURIZATION

The Shell residual oil hydrodesulfurization process is broadly defined as a process to improve the quality of residual oils by removing sulfur, metals, and asphaltenes as well as bringing about a reduction in the viscosity of the feedstock. The process is suitable for a wide range of the heavier feedstocks, irrespective of the composition and origin, and even includes those feedstocks that are particularly high in metals and asphaltenes.

The process centers around a fixed-bed downflow reactor which allows catalyst replacement without causing any interruption in the operation of the unit (Fig. 7-1). Feedstock is introduced to the process via a filter (backwash, automatic) after which hydrogen and recycle gas are added to the feedstock stream which is then heated to reactor temperature by means of feed-effluent heat exchangers whereupon the feed stream passes down through the reactor in trickle flow. Sulfur removal is excellent (Table 7-1), and substantial reductions in the vanadium content and asphaltenes content are also noted. In addition, a marked increase occurs in the API gravity, and the viscosity is reduced considerably.

A bunker reactor provides extra process flexibility if it is used upstream from the desulfurization reactor, especially with reference to the processing of feedstocks with a high metal content. A catalyst with a

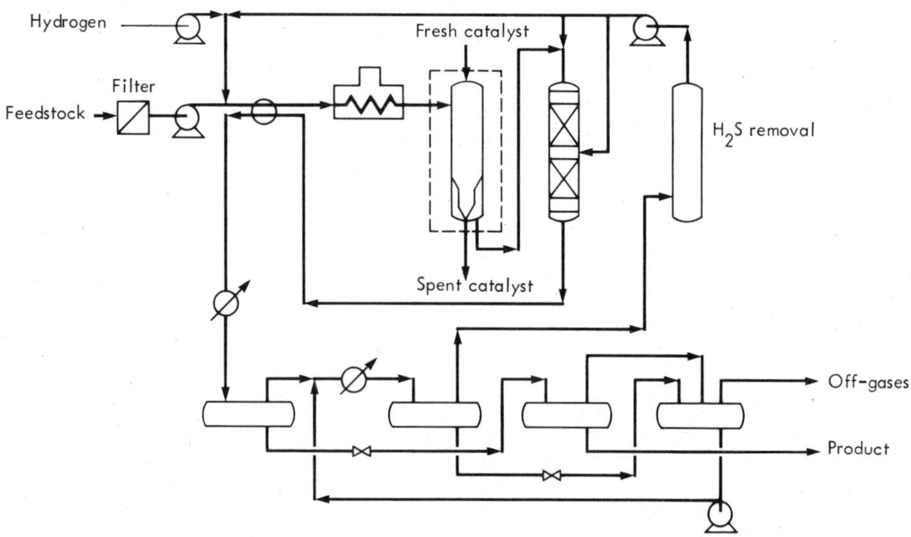

FIGURE 7-1 Flow sheet for the Shell residual oil hydrodesulfurization process indicating the upstream position of the bunker reactor.

TABLE 7-1 Operating Conditions and Product Specifications for the Shell Residual Oil Hydrodesulfurization Process as Applied to a Middle East (Kuwait) Residuum

Feedstock (350°C+; 660°F+)	
°API	14.8
Viscosity, cSt (50°C)	580
Sulfur, wt %	4.18
Vanadium, ppm	50
Asphaltenes, wt % (C_7 insolubles)	2.5
Product (165°C+; 330°F+) yields and inspections	
°API	22.8
Viscosity, cSt (50°C)	61
Sulfur, wt %	0.50
Vanadium, ppm	12
Asphaltenes, wt % (C_7 insolubles)	1.2
Yields, wt %	
C_5-165°C (330° F)	1.0
165° C+ (330° F+)	95.4

capacity for metals is employed in the bunker reactor to protect the desulfurization catalyst from poisoning by the metals. In the case of the bunker reactor (Fig. 7-2), fresh catalyst is loaded into a storage vessel (V-1) at atmospheric pressure from which it passes into a sluice vessel (V-2) where pressurization occurs prior to passage to the bunker reactor. In the reactor, inverted cone segments support the catalyst and are designed to allow catalyst removal. Thus, catalyst is withdrawn from the reactor into the sluice vessel (V-3) which is then depressurized and the contents transferred to a discharge vessel (V-4).

7-2 THE H-OIL PROCESS

The H-Oil process can be used to desulfurize and demetallize residua as well as to produce low-sulfur coker feedstocks in addition to hydrocracking residua and heavy oils into lighter liquid products. Thus, the process has a flexibility that allows response to market demands where any given H-Oil installation can vary from a predominantly hydrodesulfurization operation to a conversion (hydrocracking) operation.

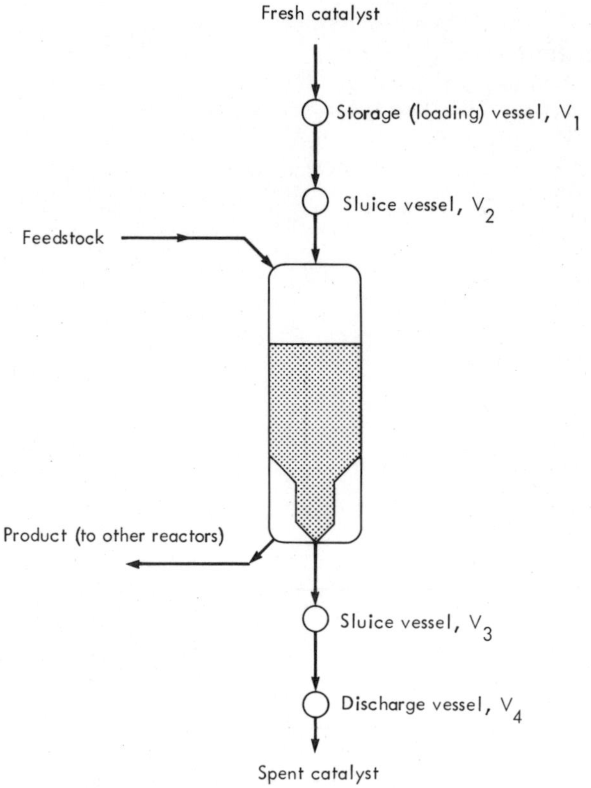

FIGURE 7-2 Schematic representation of a bunker reactor for the Shell residual oil hydrodesulfurization process.

The H-Oil process (Fig. 7-3) employs an ebullated-bed reactor in which feedstock and hydrogen pass upward through the catalyst bed. High linear velocities expand the catalyst bed in excess of its settled volume, causing movement of the individual catalyst particles, thereby ensuring distribution of feedstock and hydrogen throughout the catalyst bed and improving the efficiency of the process. This type of catalyst system also allows continuous addition and withdrawal of the catalyst, thereby eliminating the necessity for frequent shutdowns for catalyst regeneration, and it also allows process operation at conditions nearer to the limits of the catalyst system.

The ebullated reactor is usually a single-stage unit (Fig. 7-4) although two-stage units with liquid recycle (Fig. 7-5) have also been recommended.

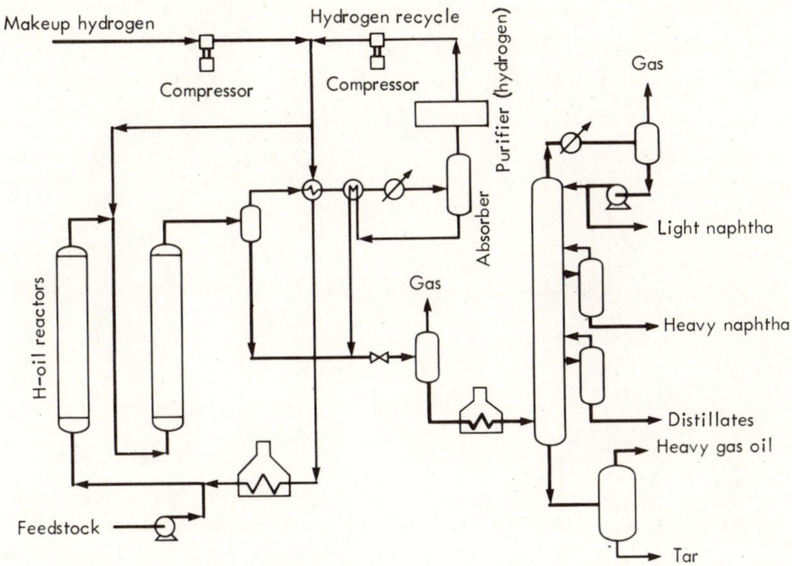

FIGURE 7-3 The H-Oil process.

This latter case allows selective withdrawal of catalyst from the base of the inlet bed to minimize the deteriorating effect of metals deposition from the feedstock onto the catalyst.

The H-Oil process can be used for most residua (Tables 7-2, 7-3) as well as for materials such as bitumen extracted from the tar sands located in northeastern Alberta, Canada (Table 7-4). In the latter case, distillate hydrotreating is recommended for inclusion within the process (Fig. 7-6) to produce a high-quality synthetic crude oil by making maximum use of the available temperature and the pressurized hydrogen. This produces low-sulfur liquid streams as well as a naphtha fraction that is essentially free of nitrogen.

7-3 THE GULF RESID HYDRODESULFURIZATION PROCESS

This process is suitable for the desulfurization of high-sulfur residua (atmospheric and vacuum) to produce low-sulfur fuel oils or catalytic cracking feedstocks. In addition, the process can be used, through alternate design types, to upgrade high-sulfur crude oils or bitumens that are unsuited for the more conventional refining techniques.

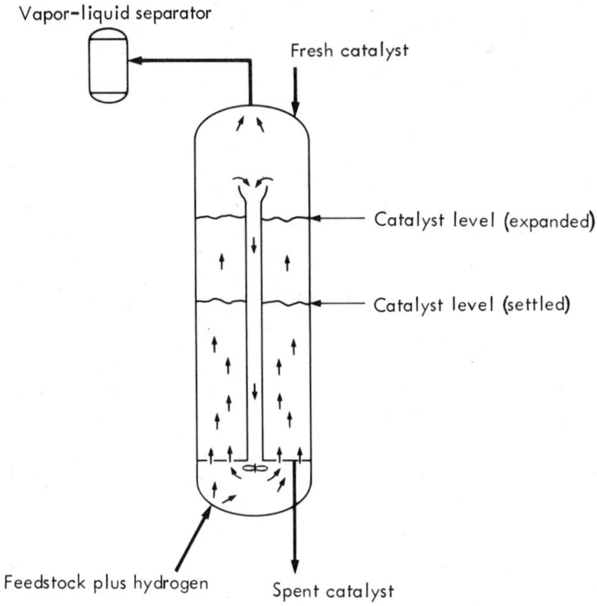

FIGURE 7-4 Schematic representation of a single-stage H-Oil reactor.

TABLE 7-2 H-Oil Processing of a Californian Vacuum Residuum

Feedstock:			
°API		8.8	
Sulfur, wt %		2.01	
Nitrogen, wt %		0.91	
Ramsbottom carbon, wt %		14.6	
Percent boiling to 975°F		28.3	
Percent boiling above 975°F		71.7	
Product yields and inspections	**Vol %**	**°API**	**Sulfur (wt %)**
C_4–200°F	10	90	–
200–400°F	23	54	0.1
400–500°F	17	36	0.2
500–650°F	23	28	0.5
650–975°F	24	14	1.0
975°F+	11	-10	2.0
C_4–975°F+	108	27	0.6

TABLE 7-3 H-Oil Processing of a Kuwait Atmospheric Residuum with (a) 40% Desulfurization and (b) 80% Desulfurization

Feedstock	
°API	14.0
Sulfur, wt %	4.0
Viscosity	
SFS 122°F	605
SFS 210°F	43
Ramsbottom carbon, wt %	9.2
Percent boiling to 975°F	50.0
Percent boiling above 975°F	50.0

	Vol %	°API	Sulfur (wt %)	(SFS at 122°F)
(a) Product yields and inspections:				
C_5-400°F	1.8	57.0	–	–
400-680°F	12.8	32.0	0.7	–
680°F+	86.6	16.5	2.5	170
C_5-680°F+	101.2	–	–	
Hydrogen consumption: 300 scf/bbl				
(b) Product yields and inspections:				
C_5-400°F	5.0	57.0	0.1	–
400-680°F	15.0	32.0	0.3	–
680°F+	81.0	17.5	1.0	75
C_5-680°F+	102.0	–	–	–
Hydrogen consumption: 600 scf/bbl				

The process has three basic variations—the Type II unit, the Type III unit, and the Type IV unit with the degree of desulfurization, and process severity, increasing from Type II to Type IV. Thus, liquid products from Types III and IV units can be used directly as catalytic cracker feedstocks and perform similarly to virgin gas oil fractions, whereas liquid products from the Type II unit usually need to be vacuum-flashed to provide a feedstock suitable for a catalytic cracker.

Fresh, filtered feedstock is heated together with hydrogen and recycle gas and charged to the downflow reactor from which the liquid product goes

TABLE 7-4 H-Oil Processing of Athabasca Bitumen

Feedstock			
°API		8.3	
Sulfur, wt %		4.9	
Nitrogen, wt %		0.45	
Percent boiling to 975°F		49.7	
Percent boiling above 975°F		50.3	

Product yields and inspections	Vol %	°API	Sulfur (wt %)
C_5-380° F	16.0	51.0	1.0
380-650°F	43.0	28.0	2.0
650-975°F	26.4	13.0	3.5
975° F+	16.0	-9.0	5.7
C_5-975° F	101.4	21.2	2.7

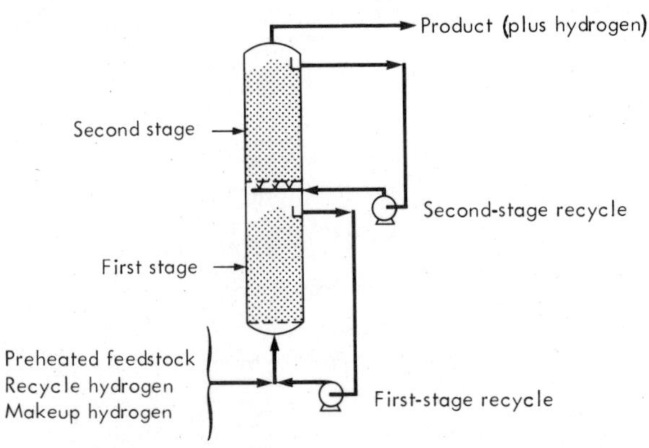

FIGURE 7-5 Schematic representation of a two-stage H-Oil reactor with liquid recycle to each stage.

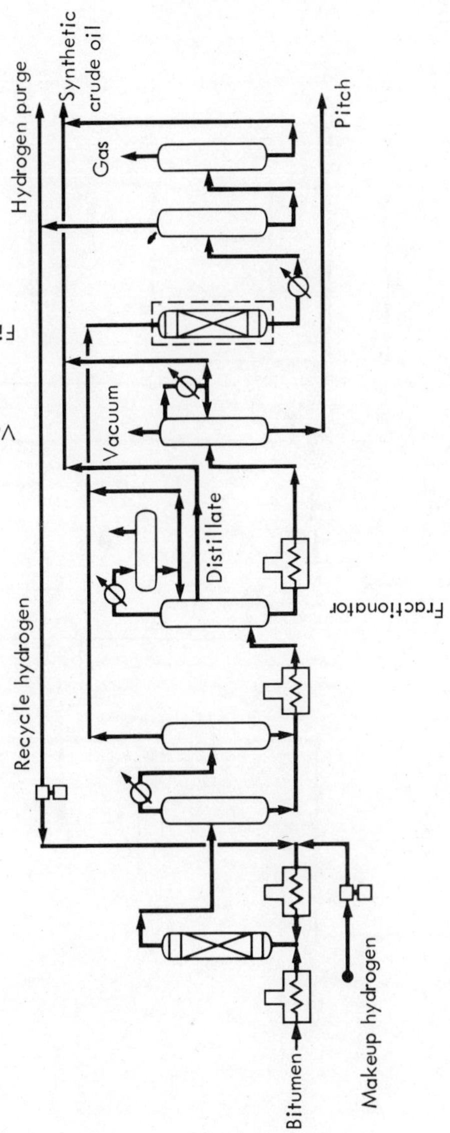

FIGURE 7-6 H-Oil processing of Athabasca bitumen with distillate hydrotreating.

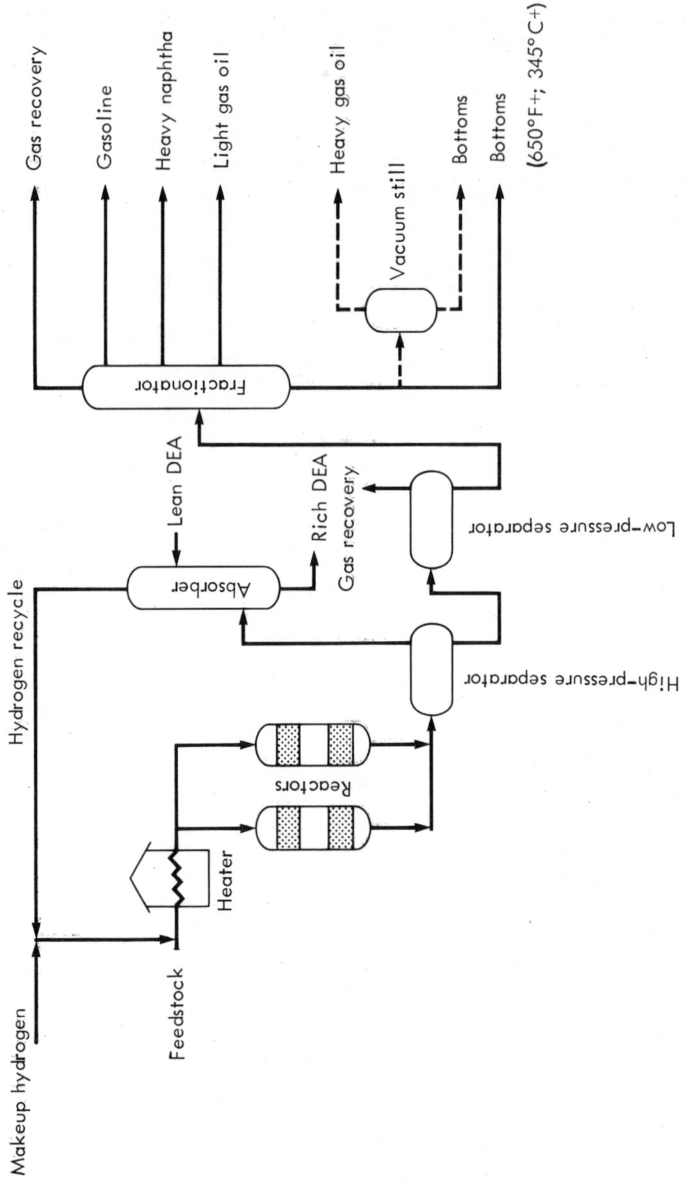

FIGURE 7-7 The Gulf Resid Hydrodesulfurization process.

to fractionation after flashing to produce the various product streams. Each process type is basically similar to its predecessor (Fig. 7-7) but will differ in the number of reactors. For example, modifications necessary to convert the Type II to the Type III process consist of the addition of a reactor and related equipment, while the Type III process can be modified to a Type IV process by the addition of a third reactor section. Types III and IV are especially pertinent to the problem of desulfurizing heavy oils and residua since they have the capability of producing extremely low-sulfur liquids from high-sulfur residua (Table 7-5).

7-4 THE UNICRACKING HYDRODESULFURIZATION PROCESS (UNION OIL COMPANY)

The Unicracking Hydrodesulfurization process is a fixed-bed catalytic process that is designed to remove sulfur from atmospheric and vacuum residua and is often considered as a true hydrodesulfurization process inasmuch as conversion of the feedstock to lighter products is minimized.

In this process (Fig. 7-8), the feedstock and hydrogen (usually hydrogen-enriched gas) are preheated, mixed, and introduced into a guard chamber containing the process catalyst which effectively removes any mineral matter (particulates) and residual salt remaining in the feed. The cleaned feedstock then enters the reactor(s) where it contacts the major portion of the catalyst in a downflow operation. Hydrogen consumption is minimized because of the relatively low pressures and the tendency to control the amount of hydrocracking and saturation of aromatic entities. Products from the reactor(s) are cooled and separated from recycle gas to give a low-sulfur fuel oil that represents a substantial portion of the feedstock (Table 7-6).

The process is claimed to operate satisfactorily for a variety of feedstocks varying in sulfur content from about 1.0% by weight to about 5% by weight although the rate of desulfurization is dependent on the sulfur content of the feedstock (Table 7-7, Fig. 7-9) as is catalyst life (Fig. 7-10), product sulfur (Fig. 7-11), and hydrogen consumption (Fig. 7-12).

The high efficiency of the process is believed to be due to the excellent distribution of the feedstock and hydrogen which occurs in the reactor where a proprietary liquid distribution system is employed (Fig. 7-13). In addition, the process catalyst (also proprietary) was designed for the desulfurization of residua and is not merely an upgraded gas oil hydrotreating catalyst as often occurs in various processes. It is a cobalt-molybdenum-alumina catalyst with a controlled pore structure that permits a high degree of desulfurization and, at the same time, minimizes any coking tendencies (Fig. 7-14).

TABLE 7-5 Process Data for the Gulf Resid Hydrodesulfurization Units

Feedstock specifications	South Louisiana	West Texas	Light Arabian	Kuwait
°API	22.4	19.3	18.5	16.6
Sulfur, wt %	0.46	2.2	2.93	3.8
Nitrogen, wt %	0.16	0.20	0.16	0.21
Carbon residue (Ramsbottom), wt %	1.76	5.5	6.79	8.3
Nickel, ppm	3.2	7.1	7.3	15
Vanadium, ppm	1.7	14.0	27.0	45
ASTM distillation, °F				
10%	681	676	640	686
30%	751	767	764	809
50%	825	882	873	950
70%	912	–	–	–
90%	1060	–	–	–
Yield on Crude, vol %	48	44.1	41	53

Product specifications	South Louisiana	West Texas	Light Arabian	Kuwait	
Hydrodesulfurization type	II	II	IV	III	IV
Desulfurization, %	87	85	95.9	92	97
°API	24.0	23.8	25.1	25.7	26.0
Sulfur, wt %	0.05	0.33	0.12	0.28	0.10
Nitrogen, wt %	0.10	0.14	0.09	0.14	0.11
Carbon residue (Ramsbottom), wt %	1.4	2.21	2.28	2.83	2.2
Nickel, ppm	0.1	1.1	0.5	1.1	0.2
Vanadium, ppm	0.1	0.7	0.3	0.8	0.1
Demetallization, %	97.1	87.5	95.3	94.6	99.2
ASTM distillation, °F					
10%	681	660	611	574	540
30%	772	756	735	707	723
50%	848	850	829	816	831
70%	944	–	947	940	963
90%	1080	–	1075	–	–
Yield on whole crude, vol %	47.7	43.8	39.8	52.0	51.7
Yield on feedstock, vol %	99.4	99.3	97.0	98.1	97.5

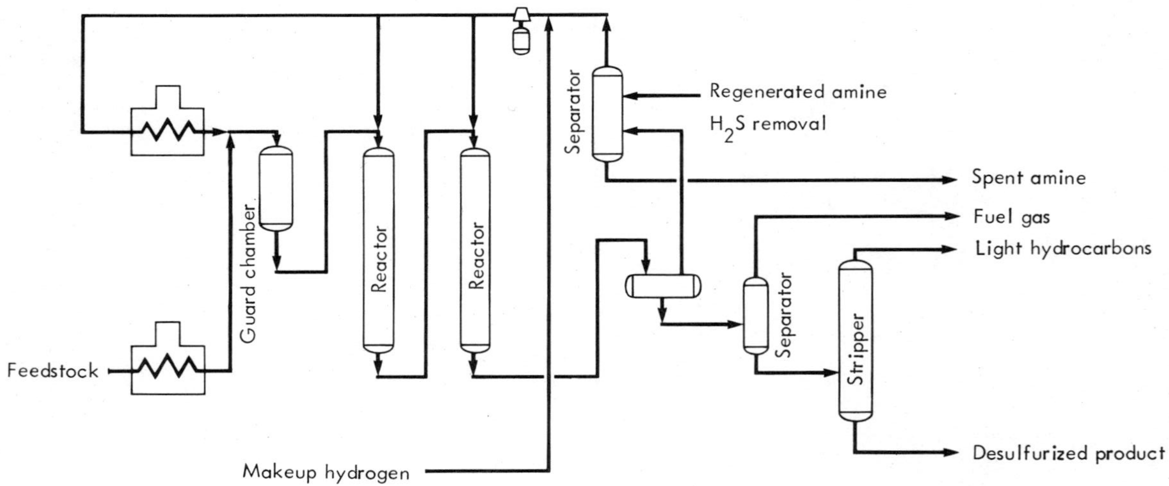

FIGURE 7-8 Flow sheet for the Union Oil Unicracking/Hydrodesulfurization process.

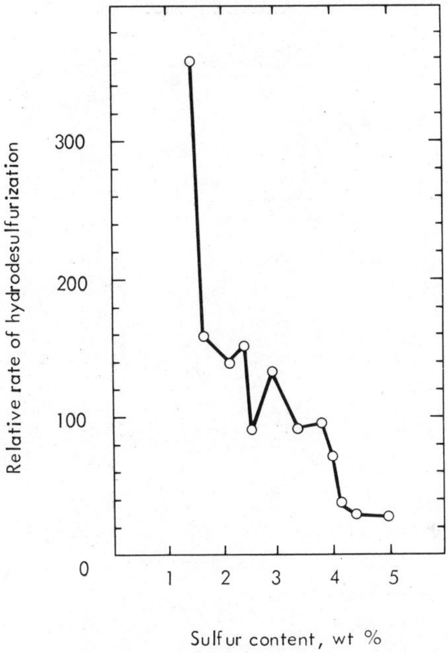

FIGURE 7-9 Variation of desulfurization rate with sulfur content of the feedstock in the Unicracking/Hydrodesulfurization process (see Table 7-7). (Reproduced from M. Skripek, S. K. Alley, and P. F. Neuth, Proceedings of the 79th National Meeting of the American Institute of Chemical Engineers, Houston, Texas, March 1978, with permission of the publisher.)

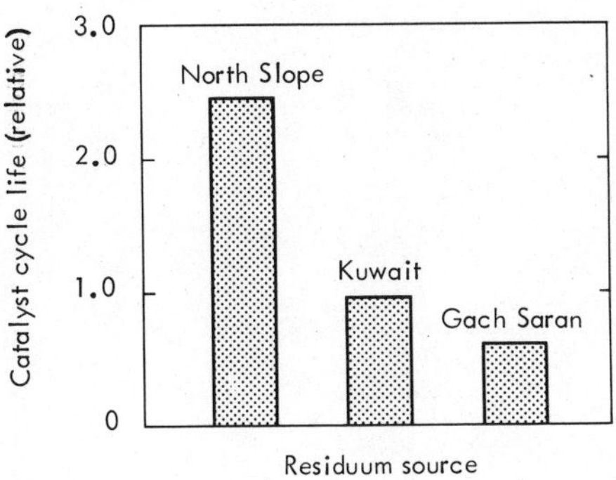

FIGURE 7-10 Variation of catalyst life to produce a constant sulfur content in the product (see Table 7-7).

TABLE 7-6 Hydrodesulfurization of Various Feedstocks by the Unicracking/HDS Process

Feedstock	Kuwait (650°F+)	Gach Saran (650°F+)	North Slope (650°F+)
°API	16.7	16.3	15.2
Sulfur, wt %	3.8	2.4	1.65
Nitrogen, wt %	0.21	0.44	0.37
Conradson carbon, wt %	8.5	8.5	8.4
Ni + V, ppm	46	220	44
Product specifications			
C_5-365°F	1.1	1.8	0.8
365°F			
Yield, vol %	100.4	100.4	100.5
°API	24.4	22.8	19.8
Sulfur, wt %	0.30	0.30	0.30
Nitrogen, wt %	0.11	0.27	0.26
Conradson carbon, wt %	3.0	4.0	4.0
Ni + V, ppm	15	55	14
H_2 consumption, scf/bbl	730	570	340

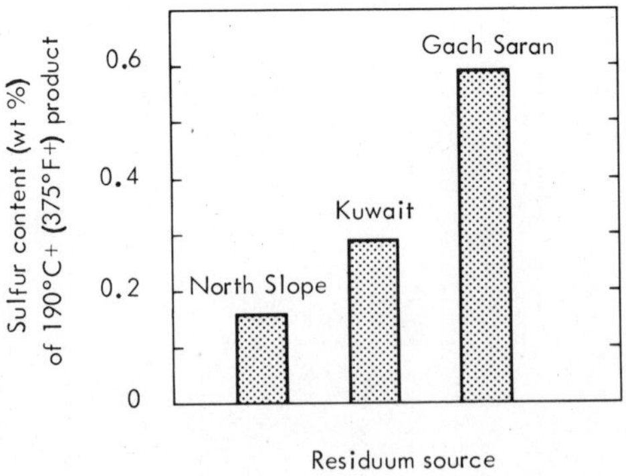

FIGURE 7-11 Variation of product sulfur with feedstock (see Table 7-7).

TABLE 7-7 Variation of Desulfurization Rate with Sulfur Content of the Feedstock in the Unicracking/HDS Process

Crude source	Crude (vol %)	Resid properties, 650° F+ Gravity (°API)	Sulfur (wt %)	Ni + V (ppm)	Conradson carbon (wt %)	Relative desulfurization rate
Murban	35.7	22.8	1.5	4	3.7	18.0
North Slope	52.6	15.1	1.7	45	8.9	7.75
Agha Jari (Iranian light)	42.7	17.5	2.4	98	7.7	7.25
Venezuelan (Leona)	55.0	18.9	2.1	187	8.6	7.0
Arabian light	44.7	17.9	2.9	28	7.3	6.5
L.A. Basin (Torrey Canyon)	48.4	8.4	2.0	236	11.7	4.5
West Texas Sour	41.6	15.5	3.4	29	9.0	4.5
Gach Saran (Iranian heavy)	46.7	14.6	2.2	258	9.9	4.25
Sassan	44.7	16.0	3.4	38	8.5	4.25
Kirkuk	39.3	17.6	3.7	84	8.4	4.0
Kuwait	50.5	15.5	3.9	51	8.8	3.5
Safaniyah (Arabian heavy)	53.1	12.3	4.2	123	13.3	1.5
Khafji	53.0	11.9	4.3	75	15.6	1.25
Ratawi	62.0	10.2	5.0	57	14.1	1.0

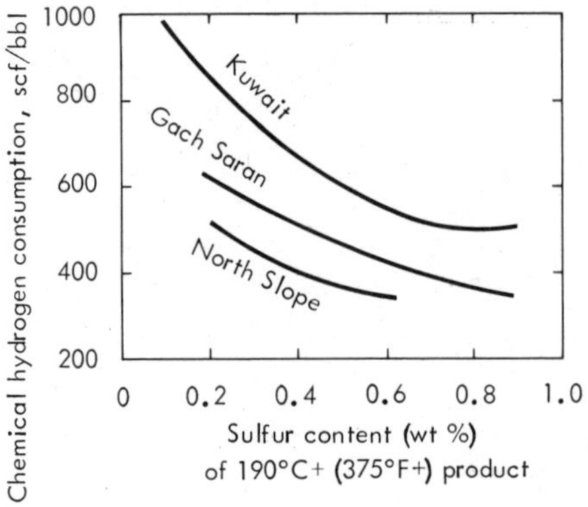

FIGURE 7-12 Variation of hydrogen consumption with product for different feedstocks (see Table 7-7).

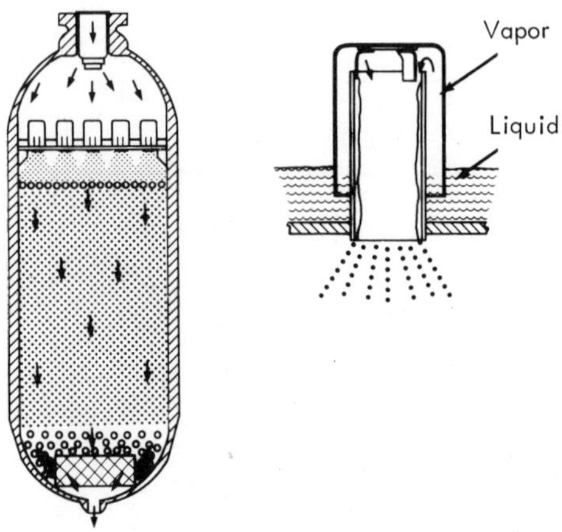

FIGURE 7-13 Liquid distribution system in the reactor for the Unicracking/Hydrodesulfurization process.

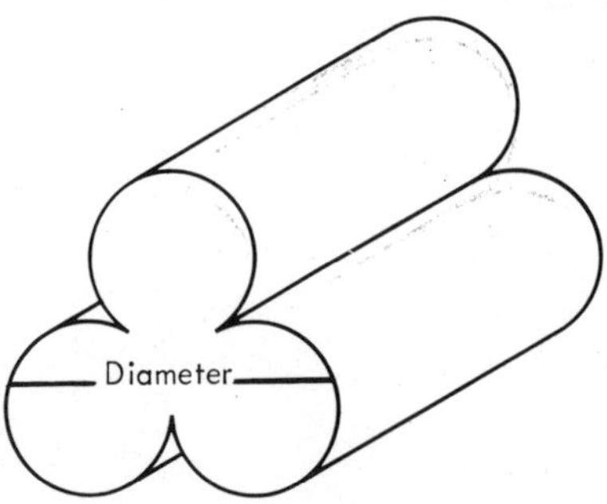

FIGURE 7-14 Cross section of the extruded catalyst for the Unicracking/Hydrodesulfurization process.

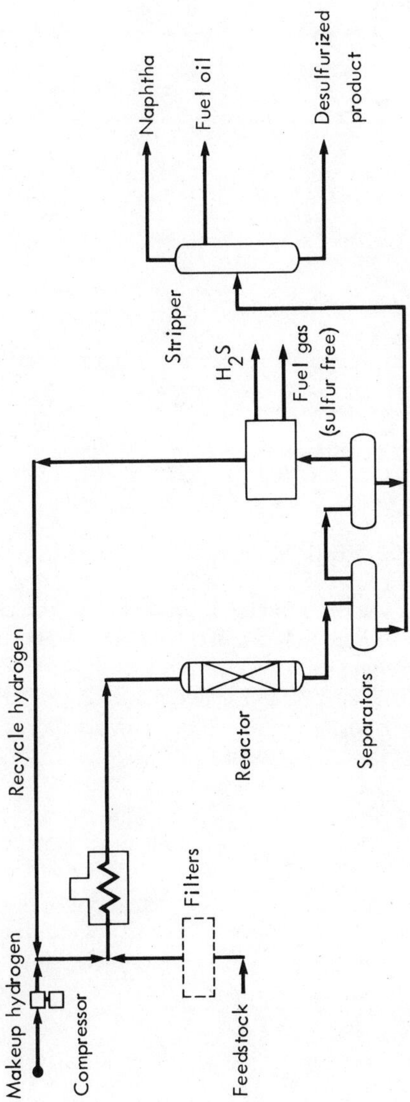

FIGURE 7-15 Flow sheet for the Exxon Residfining process.

TABLE 7-8 Feedstock and Product Specifications for the Exxon Residfining Process

Feedstock	Gach Saran (650°F+)	Arabian heavy (650°F+)
°API	15.0	12.3
Sulfur, wt %	2.50	4.19
Products		
°API	19.6	20.7
Sulfur, wt %	0.3	0.3
C_5/400°F, vol %	3.4	6.0
400°F+, vol %	98.1	96.4
Chemical H_2 consumption, scf/bbl		

7-5 THE EXXON RESIDFINING PROCESS

The Exxon Residfining process is designed to desulfurize atmospheric residua and to produce a liquid product that is of the heavy fuel oil type and that has a sulfur content substantially lower than the feedstock. In the process (Fig. 7-15), the feedstock and hydrogen-rich gas are preheated prior to entering the guard chamber which acts as a reactor plugging protection system to protect the fixed-bed downflow reactor from the coking and fouling tendencies of the heavier feedstocks. The reactor effluent is then cooled, separated from the hydrogen and light ends to yield a desulfurized product that has a marked improvement in properties over the feedstock (Table 7-8).

The reactors use proprietary catalysts that maintain excellent activity with residua feedstocks and have adequate life at low pressures—a necessary property of the catalyst under the circumstances since the process is, in essence, a low conversion process with operating pressures of about 1000 psi.

7-6 THE LC-FINING PROCESS (C-E LUMMUS AND CITIES SERVICE RESEARCH AND DEVELOPMENT COMPANY)

The LC-Fining process is designed to produce low-sulfur liquids from residua or other fossil fuel bases such as coal liquids, coal tars, or shale oils. With respect to petroleum residua, the process can operate with

TABLE 7-9 Feedstock and Product Specifications for the LC-Fining Process

Feedstock	Kuwait atmospheric residuum	Gach Saran vacuum residuum	Arabian heavy vacuum residuum
°API	15.0	6.1	7.5
Sulfur, wt %	4.05	3.52	4.89
Products			
Yield, wt %			
$C_3 + C_4$	0.8	1.6	2.1
C_5-400° F	2.5	9.7	14.3
400-650°F	22.7	14.1	26.5
650-975°F	34.7	24.1	31.1
975°F+	35.5	47.5	21.3
Total liquids			
°API	23.0	21.2	25.5
Sulfur, wt %	0.4	0.9	1.22
Hydrogen consumption, scf/bbl	690	1400	1410

feedstocks which have high asphaltene contents or high metal contents without the need for shutdowns to replace the catalyst. In addition, hydrodesulfurization or hydrocracking can be achieved by changing the severity of the process conditions.

The process (Fig. 7-16) involves upflow of the heated liquid feedstock and hydrogen through an expanded bed of catalyst in a suitably constructed reactor (Fig. 7-17). The turbulence of the catalyst bed (which is provided by the upward velocity of the reactants) provides efficient contact between the reactants and maintains isothermal conditions. The rapid, highly exothermic, reaction has no need for an internal quench system in the reactor since the incoming feedstock can serve as a quench. Thus, it is possible to produce low-sulfur liquid products from various heavy feedstocks (Table 7-9) with only minimal conversion to light products, thereby maintaining a low level of hydrogen consumption.

The process continuity is maintained by intermittent addition and withdrawal of the catalyst addition rate. In a similar manner, it is possible to induce greater degrees of hydrocracking by adjustment of the temperature to the upper limits of the process.

With regard to the truly heavy feedstocks, the LC-Fining process has been applied to desulfurization of bitumen extracted from the Athabasca tar

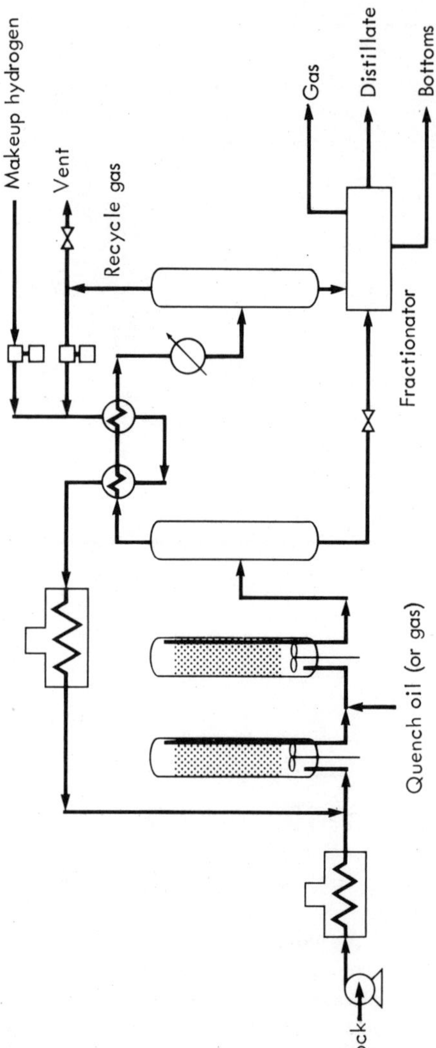

FIGURE 7-16 Flow sheet for the LC-Fining process.

TABLE 7-10 Feedstock and Product Specifications for Application of the LC-Fining Process to Athabasca Bitumen

	Bitumen type	
Feedstock	High solids	Low solids
°API	5.5	9.1
Sulfur, wt %	4.9	4.6
Nitrogen, wt %	0.4	0.4
Ash, wt %	3.2	0.8
Distillation, vol %		
C_5-650° F	18	17.3
650-975°F	36	37.3
975°F+	46	45.4
Gravity, °API	-4.3	0.0
Sulfur, wt %	6.4	6.1

Product specifications	wt %	vol %	wt %	vol %
C_1-C_4	4.8	–	5.2	–
C_5-400° F	11.7	15.5	11.9	15.5
400-650°F	32.5	37.5	37.7	43.2
650-975°F	35.5	36.8	30.0	31.7
975°F+	14.1	11.5	12.9	11.4
Desulfurization, wt %		44		77
Chemical hydrogen consumption, scf/bbl		700		1100
Total liquid				
°API		17.6		24.3
Sulfur, wt %		2.9		1.1
C_5-400° F				
°API		50		51
Sulfur		1.0		0.2
400-650° F				
°API		26.9		29.5
Sulfur, wt %		2.1		0.7
650-975° F				
°API		10.4		16.8
Sulfur, wt %		3.3		1.1
975° F+				
°API		20.0		-7.0
Sulfur, wt %		5.4		3.4

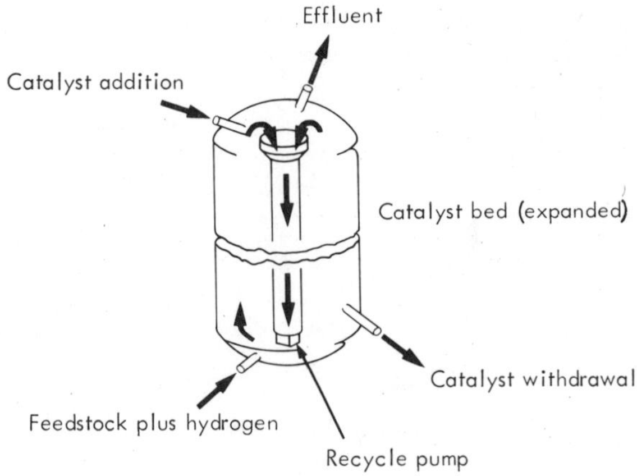

FIGURE 7-17 An expanded-bed reactor as used in the LC-Fining process.

sands. In one reported instance, a low-solids bitumen and a high-solids bitumen were employed as feedstocks whereupon good conversion of the bitumen to lower-sulfur products was noted (Table 7-10). However, the process was not claimed to be the means by which conversion of the bitumen to a synthetic crude oil would be achieved; as with most bitumen conversion processes, the LC-Fining process would play the part of the primary conversion process from which liquid products would accrue—these products would then pass to a secondary upgrading (hydrotreating) process to yield a synthetic crude oil.

7-7 THE CHEVRON RDS ISOMAX AND VRDS PROCESSES

These processes are designed to remove sulfur, nitrogen, asphaltene, and metal contaminants from residual stocks and are also capable of accepting whole crude oils or topped crude oils as feedstocks. The major product of the processes is a low-sulfur fuel oil and the amount of gasoline and middle distillates is maintained at a minimum to conserve hydrogen.

The basic elements of each process are similar (Fig. 7-18) and consist of a once-through operation of the feedstock coming into contact with hydrogen and the catalyst in a downflow reactor which is designed to maintain activity and selectivity in the presence of deposited metals. Moderate temperatures and pressures are employed to reduce the incidence of hydrocracking and, hence, minimize production of low-boiling distillates (Table 7-11).

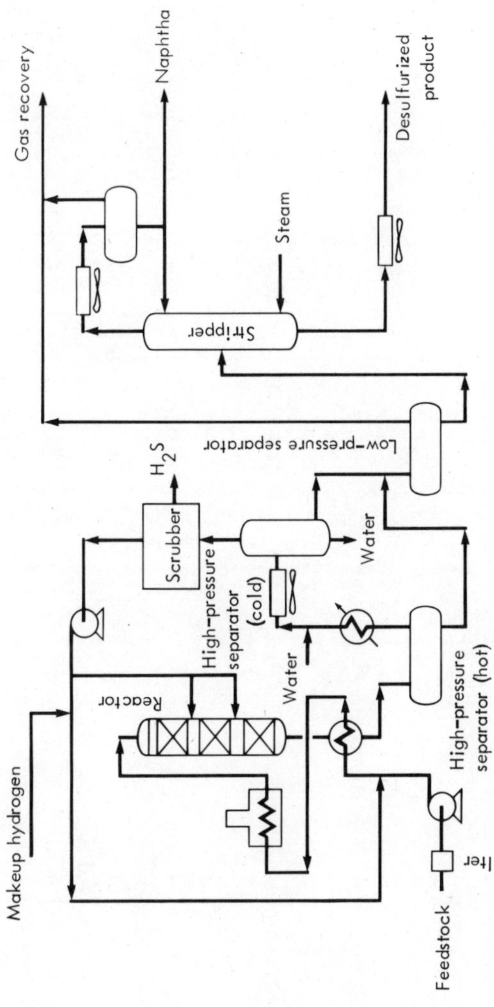

FIGURE 7-18 Flow sheet for the Chevron RDS (and VRDS) processes.

TABLE 7-11 Feedstock and Product Specifications for the Chevron RDS and VRDS Processes

	Arabian light		Arabian heavy	
	Atmospheric	Vacuum	Atmospheric	Vacuum
Feed properties				
Gravity, °API	17.7	6.5	16.8	6.1
Sulfur, wt %	3.2	4.1	3.9	5.1
Distillation, vol % boiling below				
650°F	5	—	16	—
1000°F	67	1	60	10
H_2 consumption (chemical), scf/bbl	560	750	780	960
Product yields				
$\leqslant C_4$, wt %	3.5	—	4.7	—
C_5-350°F, vol %	1.4	5.7	1.4	6.4
350-650°F, vol %	11.1		25.0	
650-1000°F, vol %	64.4	18.1	43.4	26.6
1000°F+, vol %	24.3	78.4	31.4	69.9
Sulfur content, wt %				
650°F+	0.46	1.0	0.55	1.0
1000°F+	0.97	1.2	0.88	1.25

The combination of a vacuum gas oil desulfurization step and a vacuum residuum desulfurizer (VRDS) is often seen as an attractive alternate to the atmospheric residuum desulfurizer (RDS) because the combination route uses less hydrogen for a similar investment cost. Nevertheless, the alternate processes must be evaluated on the basis of the feedstock as well as the desired product quality which strongly influence the economics of the processes.

7-8 THE CHEVRON DEASPHALTED OIL HYDROTREATING PROCESS

The Chevron deasphalted oil hydrotreating process is designed to desulfurize heavy feedstocks which have had the asphaltene fraction removed by prior application of a deasphalting process (Chapter 4). The principal product is a low-sulfur fuel oil that can be used as a blending stock or as a feedstock for a fluid catalytic unit (Table 7-12).

The process (Fig. 7-19) employs a downflow, fixed-bed reactor containing a highly selective catalyst which provides extensive desulfurization

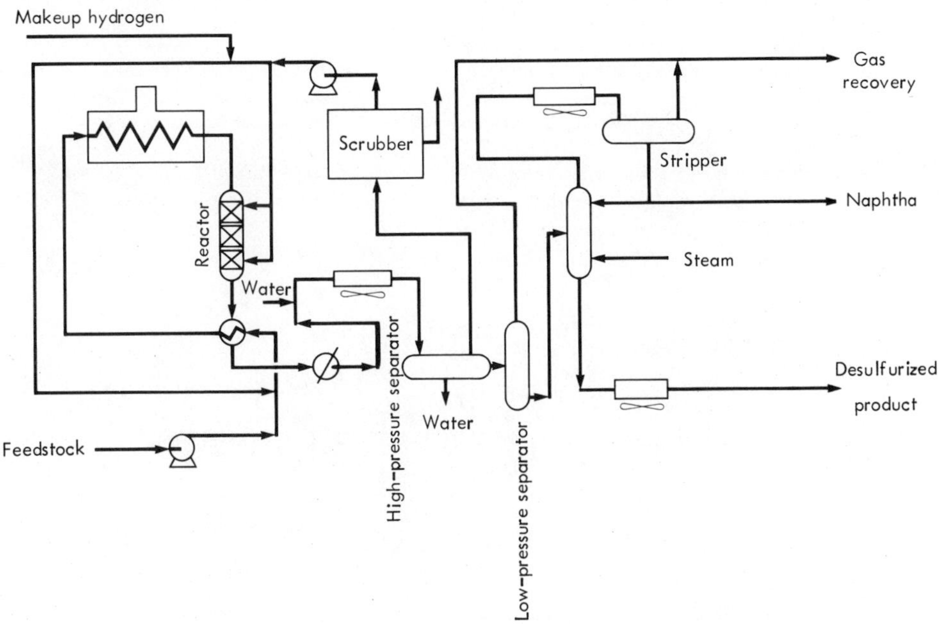

FIGURE 7-19 Flow sheet for the Chevron deasphalted oil hydrotreating process.

TABLE 7-12 Feedstock and Product Specifications for the Chevron Deasphalted Oil Hydrotreating Process

Feedstock	
Iranian deasphalted residuum	
°API	11.4
Sulfur, wt %	2.8
Ni + V, ppm	61
Hydrogen consumption (chemical), scf/bbl	470
Product yields	
C_1-C_4, wt %	0.5
C_5-350° F, vol %	1.7
350°F+, vol %	101.3
350°F+ product	
Sulfur, wt %	0.5

at low pressures with minimal cracking and, therefore, low consumption of hydrogen.

7-9 THE RCD UNIBON PROCESS (UOP)

The RCD Unibon process is designed to produce low-sulfur fuel oils from high-sulfur atmospheric residua; the products from this process can also serve as feedstocks for coking and fluid catalytic cracking units.

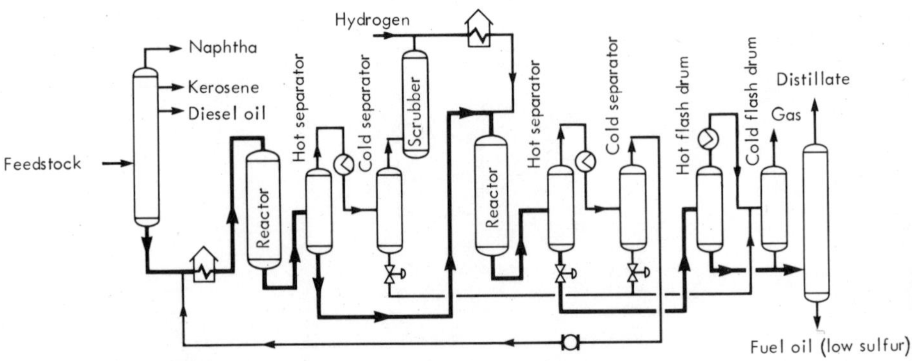

FIGURE 7-20 Flow sheet of the UOP RCD Unibon process.

TABLE 7-13 Feedstock (Kuwait Atmospheric Residuum) and Product Inspections from the UOP RCD Unibon Process

		Products	
	Feedstock	Fuel oil (375°F+)	Naphtha
Yield, vol %	–	100.9	1.0
°API	16.4	23.8	53.1
Specific gravity	0.9566	0.9109	0.7655
Viscosity, cSt at 122°F (50°C)	293	75	–
Sulfur, wt %	3.52	0.28	–
Nitrogen, wt %	0.20	0.12	–
Conradson carbon, wt %	9.45	3.84	–
Asphaltenes (C_7 insolubles), wt %	2.37	0.80	–
Pour point, °F	+55	+37	–
Hydrogen consumption, 706 scf/bbl			

The two-stage process (Fig. 7-20) employs downflow fixed-bed reactors to achieve deep desulfurization which ultimately results in the production of liquid products with substantially less sulfur than the feedstock (Table 7-13). Thus, the atmospheric residuum (from a desalted crude oil) is filtered in automatic backwash filters, combined with recycle gas, and then charged (via the heaters) to the first reactor. Desulfurization of the feedstock (to about 1 wt % sulfur) occurs by mixed-phase flow through a fixed catalyst bed in the presence of hydrogen. The first-stage effluent is combined with more hydrogen before passing to the second-stage reactor (which is operated under similar conditions to the first-stage reactor) where additional desulfurization of the residuum occurs.

Although the primary function of this process was originally to convert residua to the heavier fuel oils, the process can, nevertheless, be extended to produce distillates (varying from naphthas to vacuum gas oils) from the heavy feedstock and, at the same time, yield a diminished quantity of an upgraded vacuum residuum.

BIBLIOGRAPHY

Aalund, Leo: Hydrodesulphurisation technology takes on the sulphur challenge, Oil and Gas Journal, 79, September 11, 1972.

Bland, William F., and Davidson, Robert F. (eds.): Petroleum Processing Handbook, McGraw-Hill, New York, 1967.

Brawley, J. M.: H-Oil units attain high throughput, long runs, Oil and Gas Journal, 65, August 19, 1974.

Bridge, A. G., Scott, J. W., and Reed, E. M.: Ways to hydroprocess resids, Hydrocarbon Processing, 74, May 1975.

Henke, A. M.: Gulf's residual HDS process, Oil and Gas Journal, 97, April 6, 1970.

Hydrocarbon Processing—1978 Refining Handbook Issue, Gulf Publishing Company, Houston, Texas, 1978.

Johnson, C. A., Nongbri, G., Lehman, L. M., and Chervenak, M. C.: Conversion of bitumen and heavy crude oils to more valuable products by the H-oil process, Proc. Oil Sands of Canada-Venezuela, Can. Inst. Mining, 224, 1977.

Kubota, K., and Karner, W. M.: Commercial desulphurisation of reduced crude—a case history, Proc. Eighth World Petrol. Cong., 4:139, 1971.

Nelson, Wilbur L.: Status of residual desulphurisation, Oil and Gas Journal, 98, July 31, 1972.

Paradis, S. G., Gould, G. D., Bea, D. A., and Reed, E. M.: Isomax process for residuum and whole crude, Chem. Eng. Progress, 67(8):57, 1971.

Ranney, Maurice W.: Desulphurisation of Petroleum, Noyes Data Corporation, Park Ridge, N.J., 1975.

Residual upgrading via the Gulf HDS process, Proc. Div. Refining, Amer. Petrol. Institute, 56:384, 1977.

Rossi, W. J., Deighton, B. S., and MacDonald, A. J.: Residuum conversion with Chevron hydrotreating, Proc. Div. Refining, Amer. Petrol. Institute, 56:397, 1977.

Taylor, Herman, Jr.: To make low-sulphur resids, Hydrocarbon Processing, 86, May 1973.

Watkins, C. H., Parker, R. J., and Pharis, J. M.: Desulphurise Kuwait reduced crude, Hydrocarbon Processing, 89, May 1973.

Yanik, S. J., Frayer, J. A., Huling, G. P., and Somers, A. E.: Latest data on Gulf HDS process, Hydrocarbon Processing, 97, May 1977.

Young, B. J., and Richardson, R. L.: Resid desulphuriser a year later, Hydrocarbon Processing, 103, September 1977.

8

HYDROGEN PRODUCTION

The chemical nature of the crude oil used as the refinery feedstock has always played the major role in determining the hydrogen requirements of that refinery. For example, the lighter, more paraffinic crude oils will require somewhat less hydrogen for upgrading to, say, a gasoline product than a heavier more asphaltic crude oil (Table 8-1), and it follows that the hydrodesulfurization of heavy oils and residua (which, by definition, is a hydrogen-dependent process) needs substantial amounts of hydrogen as part of the processing requirements.

In general, considerable variation exists from one refinery to another in the balance between hydrogen produced and hydrogen consumed in the refining operations. However, what is more pertinent to the present text is the excessive amounts of hydrogen that are required for hydroprocessing operations—whether these be hydrocracking or the somewhat milder hydrotreating processes. For effective hydroprocessing, a substantial hydrogen partial pressure must be maintained in the reactor and, in order to meet this requirement, an excess of hydrogen above that actually consumed by the process must be fed to the reactor. Part of the hydrogen requirement is met by recycling a stream of hydrogen-rich gas, but the need still remains to generate hydrogen as makeup material to accommodate the process consumption of 500 to 3000 scf/bbl depending upon whether the heavy feedstock is being subjected to a predominantly hydrotreating (hydrodesulfurization) or to a predominantly hydrocracking process.

Hydrogen is generated in a refinery by the catalytic reforming process, but there may not always be the need to have a catalytic reformer as part of the refinery sequence. Nevertheless, assuming that a catalytic reformer is part of the refinery sequence, the hydrogen production from the reformer usually falls well below the amount required for hydroprocessing purposes. For example, in a 100,000-bbl/day hydrocracking refinery, assuming intensive reforming of hydrocracked gasoline, the hydrogen requirements of the refinery may still fall some 500 to 900 scf/bbl of crude charge below

TABLE 8-1 Hydrogen Contents of Various Liquid Fuels and Petroleum Fractions (%)

Crude oils	11-14
Heavy oils	9-11
Bitumens (natural asphalts)	9-12
Straight-run residua	9-12
Gasoline	13-15
Kerosene	13.5-14.5

that necessary for the hydrocracking sequences. Consequently, an "external" source of hydrogen is necessary to meet the daily hydrogen requirements of any process where the heavier feedstocks are involved.

Several processes are available for the production of the additional hydrogen that is necessary for the various heavy feedstock hydroprocessing sequences that have been outlined elsewhere (Chapter 7), and it is the purpose of the present section to present a general description of these processes. In general, most of the external hydrogen is manufactured by steam-methane reforming or by oxidation processes. Other processes such as ammonia dissociation, steam-methanol interaction, or electrolysis are also available for hydrogen production, but economic factors and feedstock availability assist in the choice between processing alternatives.

8-1 STEAM-METHANE REFORMING

Steam-methane reforming is a continuous catalytic process which involves the reaction of steam and natural gas (methane) to produce hydrogen and carbon monoxide although there may also be some carbon dioxide formation in the process:

$$CH_4 + H_2O \longrightarrow CO + 3H_2$$

$$CH_4 + 2H_2O \longrightarrow CO_2 + 4H_2$$

In the actual process (Fig. 8-1), the natural gas feedstock must first be desulfurized in order to prevent catalyst poisoning or deactivation. The desulfurization step depends upon the nature of the sulfur-containing contaminants and can vary from the more simple ambient temperature adsorption of the sulfur-containing materials on activated charcoal to a more complex high-temperature reaction with zinc oxide to catalytic hydrogenation followed by zinc oxide treatment.

Obviously, in the case of a natural gas feedstock, a simple adsorption process might suffice, while in the case of the higher-molecular-weight feedstocks, such as the naphthas (see below), a more complete desulfurization process may be necessary.

The desulfurized feedstock is then mixed with superheated steam and passed over a nickel catalyst (730 to 845°C; 1350 to 1550°F; 400 psi) to produce a mixture of hydrogen, carbon monoxide, and carbon dioxide as well as excess steam. The effluent gases are cooled (to about 370°C; 700°F) and passed through a shift converter which promotes reaction of the carbon monoxide with stream to yield carbon dioxide and more hydrogen. The shift converter may contain two beds of catalyst with interbed cooling; the combination of the two catalyst beds promotes maximum conversion of the carbon monoxide which is very essential in the event that a high-purity product is required.

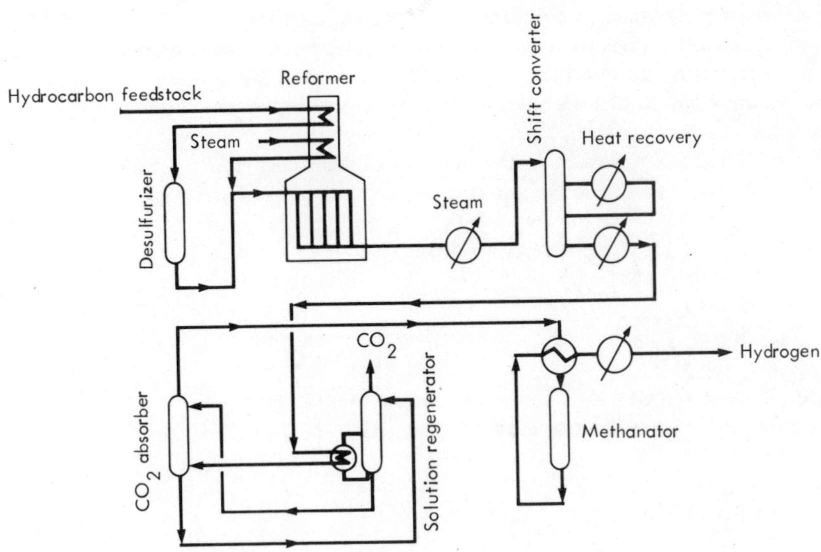

FIGURE 8-1 Steam-methane reforming.

The carbon dioxide-rich gas stream is then cooled after which carbon dioxide removal is achieved by passage through scrubbers or by the more thermally efficient hot carbonate process. Any residual carbon monoxide or carbon dioxide is removed by passing the heated gas through a nickel-base methanation catalyst where the carbon oxides are reached with hydrogen to produce methane.

8-2 STEAM-NAPHTHA REFORMING

Steam-naphtha reforming is essentially (apart from the differences in molecular weight of the feedstock) similar in nature to the steam-methane reforming process. Again, desulfurization of the feedstock is necessary as a preliminary step in the process but, as indicated above, a more severe desulfurization technique may be required. In addition, some modifications to the proprietary catalyst are necessary to accommodate the higher-molecular-weight feedstock, thereby allowing a variety of naphthas (including those with as much as 35% aromatics) boiling up to 200°C (390°F) to be used as process feedstocks.

A problem which may occur when higher-molecular-weight materials are employed as feedstocks is coke formation and deposition. When an alkali-promoted catalyst is employed, corrosion and fouling problems in the reformer (or even in equipment downstream of the reformer because of the tendency of the alkali to migrate) may occur with some frequency. However, coke formation can be eliminated by the use of a proprietary alkali-free catalyst which has an extremely high activity and resistance to poisoning.

The overall chemistry of the steam-naphtha reforming process may be represented by the following equations:

$$C_6H_{14} + 6H_2O \longrightarrow 6CO + 13H_2$$

$$C_6H_{14} + 12H_2O \longrightarrow 6CO_2 + 19H_2$$

while process details (and flowsheet) are essentially the same as those described for steam-methane reforming (Fig. 8-1).

8-3 SYNTHESIS GAS GENERATION

The synthesis gas generation process is a noncatalytic process for generating synthesis gas (principally hydrogen and carbon monoxide) for the ultimate production of high-purity hydrogen from gaseous or liquid hydrocarbons. In this process (Fig. 8-2), a controlled mixture of preheated

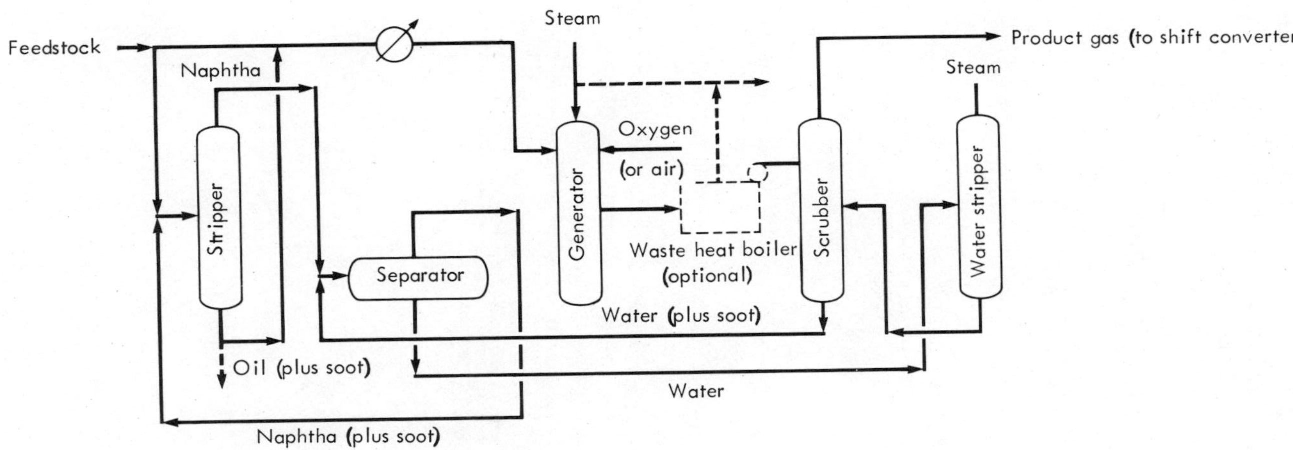

FIGURE 8-2 Synthesis gas generation by the Texaco partial oxidation process.

TABLE 8-2 Synthesis Gas Generation by the Texaco Partial Oxidation Process

Feedstock	Natural gas	Propane	64° API naphtha	9.6° API fuel oil	9.7° API fuel oil	-11.4° API coal tar
Feedstock composition, wt %						
Carbon	73.40	81.69	83.8	87.2	87.2	88.1
Hydrogen	22.76	18.31	16.2	9.9	9.9	5.7
Oxygen	-0.76	–	–	0.8	0.8	4.4
Nitrogen	3.08	–	–	0.7	0.7	0.9
Sulfur	–	–	–	1.4	1.4	0.8
Ash	–	–	–	–	–	0.1
Gross heating value, Btu/lb	22,630	21,662	20,300	18,200	18,200	15,690
Composition of Product gas, mol						
Hydrogen	61.1	54.0	51.2	45.9	45.8	38.9
Carbon monoxide	35.0	43.7	45.3	48.5	47.5	54.3
Carbon dioxide	2.6	2.1	2.7	4.6	5.7	5.7
Nitrogen	1.0	0.1	0.1	0.7	0.2	0.8
Methane	0.3	0.1	0.7	0.2	0.5	0.1
Hydrogen sulfide	–	–	–	0.1	0.3	0.2

Source: Reproduced with permission from A. H. Pelofsky (ed.), Heavy Oil Gasification, 1977, courtesy Marcel Dekker, Inc.

feedstock and oxygen is fed to the top of the generator where carbon monoxide and hydrogen emerge as the products. Soot, produced in this part of the operation, is removed in a water scrubber from the product gas stream and is then extracted from the resulting carbon-water slurry with haphtha and transferred to a fuel oil fraction. The oil-soot mixture is burned in a boiler or recycled to the generator to extinction to eliminate carbon production as part of the process.

The soot-free synthesis gas is then charged to a shift converter where the carbon monoxide reacts with steam to form additional hydrogen and carbon dioxide at the stoichiometric rate of 1 mole of hydrogen for every mole of carbon monoxide charged to the converter.

The reactor temperatures vary from 1095 to 1480°C (2000 to 2700°F), while pressures can vary from approximately atmospheric to as high as 2000 psi. The process has the capability of producing high-purity hydrogen although the extent of the purification procedure depends upon the use to which the hydrogen is to be put. For example, carbon dioxide can be removed by scrubbing with various alkaline reagents, while carbon monoxide can be removed by washing with liquid nitrogen or, if nitrogen is undesirable in the product, the carbon monoxide should be removed by washing with copper-amine solutions.

This particular partial oxidation technique has also been applied to a whole range of liquid feedstocks as a means of hydrogen production (Table 8-2), and there is now serious consideration being given to hydrogen production by the partial oxidation of solid feedstocks such as petroleum coke (from both delayed and fluid-bed reactors), lignite, and coal, as well as petroleum residua.

The chemistry of the process, using naphthalene as an example, may be simply represented as the selective removal of carbon from the hydrocarbon feedstock and further conversion of a portion of this carbon to hydrogen:

$$C_{10}H_8 + 5O_2 \longrightarrow 10CO + 4H_2$$

$$10CO + 10H_2O \longrightarrow 10CO_2 + 10H_2$$

BIBLIOGRAPHY

Bland, William F., and Robert F. Davidson, eds.: Petroleum Processing Handbook, McGraw-Hill, New York, 1967.

Hydrocarbon Processing—1975 Refining Handbook Issue, Gulf Publishing, Houston, 1975.

Pelofsky, Arnold H.: Heavy Oil Gasification, Marcel Dekker, New York, 1977.

INDEX